Mathematical and Fuzzy Methodologies for Decision Making, Management, and Engineering

This comprehensive guide bridges the gap between theoretical fuzzy logic methodologies and their practical applications in real-world scenarios. The book details the implementation of mathematical and fuzzy logic methods across decision-making, management, and engineering challenges, while addressing the growing uncertainty in modern environments that traditional methodologies fail to adequately handle.

Mathematical and Fuzzy Methodologies for Decision Making, Management, and Engineering equips readers with actionable frameworks for applying complex mathematical and fuzzy methodologies to actual business and engineering cases. The book explores real applications of diverse methodologies, introduces mathematical and fuzzy methods for decision-making, and provides step-by-step explanations of implementation procedures. Each chapter demonstrates how these advanced techniques can improve organizational outcomes through detailed case studies from multiple countries and industries, ensuring broad applicability across regional contexts.

This book is intended for graduate students, professors, researchers, and industry professionals. The new thematic trends, methodologies, and applications presented allow for techniques to be implemented with varying levels of complexity across different applications and scenarios.

Mathematical and Fuzzy Methodologies for Decision Making, Management, and Engineering

Edited by Ernesto León-Castro, Fabio Blanco-Mesa,
Victor Alfaro-Garcia, and Anna Maria Gil-Lafuente

CRC Press
Taylor & Francis Group
Boca Raton London New York

CRC Press is an imprint of the
Taylor & Francis Group, an **informa** business

Designed cover image: Getty Images

First edition published 2027
by CRC Press
2385 NW Executive Center Drive, Suite 320, Boca Raton FL 33431

and by CRC Press
4 Park Square, Milton Park, Abingdon, Oxon, OX14 4RN

CRC Press is an imprint of Taylor & Francis Group, LLC

ISBN: 978-1-041-20929-4 (hbk)
ISBN: 978-1-041-20930-0 (pbk)
ISBN: 978-1-003-72490-2 (ebk)

DOI: 10.1201/9781003724902

Typeset in Times
by KnowledgeWorks Global Ltd.

Mathematical and Fuzzy Methodologies for Decision Making, Management, and Engineering

Edited by Ernesto León-Castro, Fabio Blanco-Mesa, Victor Alfaro-Garcia, and Anna Maria Gil-Lafuente

CRC Press
Taylor & Francis Group
Boca Raton London New York

CRC Press is an imprint of the
Taylor & Francis Group, an **informa** business

Designed cover image: Getty Images

First edition published 2027
by CRC Press
2385 NW Executive Center Drive, Suite 320, Boca Raton FL 33431

and by CRC Press
4 Park Square, Milton Park, Abingdon, Oxon, OX14 4RN

CRC Press is an imprint of Taylor & Francis Group, LLC

ISBN: 978-1-041-20929-4 (hbk)
ISBN: 978-1-041-20930-0 (pbk)
ISBN: 978-1-003-72490-2 (ebk)

DOI: 10.1201/9781003724902

Typeset in Times
by KnowledgeWorks Global Ltd.

Contents

About the Editors

Ernesto León-Castro is an associate professor at the Universidad Católica de la Santísima Concepcion (Chile). He has a PhD in Administrative Sciences from Universidad de Occidente (Mexico). He is a member of the National System of Researchers (Level 2). His research areas include aggregation operators, decision-making, innovation, finance, fuzzy logic, and other related topics. He is part of the Ibero-American Network for Competitiveness, Innovation and Development (REDCID) and is a member of the Business Studies in Uncertain Environments research group.

Fabio Raul Blanco Mesa is a full professor and senior researcher at the Universidad Pedagógica y Tecnológica de Colombia. He received his PhD in Business from the Department of Economics and Business Organization of the University of Barcelona and has Master's degrees in Business, Finance and Insurance Research, and Sports Business Management from the University of Barcelona. He also has a Sports Administrator degree from Francisco José de Caldas District University. His areas of interest include issues related to decision-making, competitiveness, interest groups, entrepreneurship, regional development, strategic analysis, diffuse systems and aggregation operators applied to management, uncertainty, and bibliometrics.

Victor Alfaro-Garcia is the Deputy Coordinator of the Ibero-American Network for Competitiveness Innovation and Development REDCID, a program attached to the Ibero-American Program of Science and Technology for Development CYTED. He received his PhD in International Business with specialization in models for the treatment of uncertainty in decision-making.

Anna Maria Gil-Lafuente is a full professor in the Department of Business Administration at the University of Barcelona, Spain. She has a MSc and a PhD degree in Business Administration from the University of Barcelona. She is on the editorial board of several journals and her areas of interest include decision-making, aggregation operators, fuzzy systems, uncertainty, and finance.

Contributors

Victor Alfaro-Garcia
Universidad Michoacana de San Nicolás de Hidalgo
Morelia, Michoacan, Mexico

Pavel Anselmo Alvarez-Carrillo
Universidad Autonoma de Occidente
Culiacan, Sinaloa, Mexico

Ruhit Bardhan
Indian Institute of Teacher Education
Gandhinagar, Gujarat, India

María del Refugio Bernal Agramón
Universidad Tecnologica de Culiacan
Culiacan, Sinaloa, Mexico

Fabio Blanco-Mesa
Universidad Pedagogica y Tecnologica de Colombia
Tunja, Colombia

Dharmik Chauhan
Pandit Deendayal Energy University
Gandhinagar, Gujarat, India

David Luviano Cruz
Universidad Autónoma de Ciudad Juárez
Ciudad Juarez, Chihuahua, Mexico

Enrique Cruz-Domínguez
Universidad Autonoma Benito Juarez de Oaxaca
Oaxaca, Mexico

Ajoy Kanti Das
Tripura University
Suryamaninagar, Tripura, India

Luis Asunción Pérez Domínguez
Universidad Autónoma de Ciudad Juárez
Ciudad Juarez, Chihuahua, Mexico

Dalia García-Orozco
Universidad Michoacana de San Nicolás de Hidalgo
Morelia, Michoacan, Mexico

Diego Alonso Gastélum-Chavira
Universidad Autónoma de Occidente
Culiacan, Sinaloa, Mexico

Ana Laura Herrera-Prado
Universidad Autónoma de Occidente
Mazatlan, Sinaloa, Mexico

Alejandra Holguín-Ávila
Universidad Autónoma de Ciudad Juárez
Ciudad Juarez, Chihuahua, Mexico

Martín Isimayrt Huesca-Gastélum
Universidad Tecnologica de Culiacan
Culiacan, Sinaloa, Mexico

Ernesto León-Castro
Universidad Catolica de la Santisima Concepcion
Concepcion, Biobio, Chile

Darshana B. Likhada
Indian Institute of Teacher Education
Gandhinagar, Gujarat, India

Roberto Romero López
Universidad Autónoma de Ciudad Juárez
Ciudad Juarez, Chihuahua, Mexico

Lydia María López-Barraza
Universidad Autónoma de Occidente
Culiacan, Sinaloa, Mexico

María Guadalupe López-Barraza
Universidad Autónoma de Sinaloa
Culiacan, Sinaloa, Mexico

Pavel Lopez-Parra
Universidad Autonoma de Occidente
Culiacan, Sinaloa, Mexico

Luis Carlos Méndez-González
Universidad Autónoma de Ciudad Juarez
Ciudad Juarez, Chihuahua, Mexico

Eva Luz Miranda-Espinoza
Universidad Autonoma de Occidente
Culiacan, Sinaloa, Mexico

Victor Manuel Morales Rocha
Universidad Autónoma de Ciudad Juarez
Ciudad Juarez, Chihuahua, Mexico

Manuel Muñoz-Palma
Universidad de Sonora
Hermosillo, Sonora, Mexico

Elizabeth Olmos-Martínez
Universidad Autónoma de Occidente
Mazatlan, Sinaloa, Mexico

Patricia Cristina Parroquin-Amaya
Universidad Autónoma de Ciudad Juarez
Ciudad Juarez, Chihuahua, Mexico

Iván Juan Carlos Pérez-Olguín
Universidad Autónoma de Ciudad Juarez
Ciudad Juarez, Chihuahua, Mexico

Ana Luz Ramos-Soto
Universidad Autonoma Benito Juarez de Oaxaca
Oaxaca, Mexico

Manoj Sahni
Pandit Deendayal Energy University
Gandhinagar, Gujarat, India

Alanda Catalina Torres-Salazar
Universidad Autónoma de Sinaloa
Culiacan, Sinaloa, Mexico

Brenda Dennís Valadez-Solana
Universidad Autonoma Benito Juarez de Oaxaca
Oaxaca, Mexico

Indira C. Ventura-Ruiz
Universidad Michoacana de San Nicolás de Hidalgo
Morelia, Michoacan, Mexico

Mónica Velarde-Valdez
Universidad Autónoma de Occidente
Mazatlan, Sinaloa, Mexico

Margarita Guadalupe Zazueta-Hernández
Universidad Autónoma de Sinaloa
Culiacan, Sinaloa, Mexico

Introduction

Ernesto León-Castro, Fabio Raul Blanco Mesa, Victor Alfaro-Garcia, and Anna Maria Gil-Lafuente

This book is a compilation of eleven chapters that clearly illustrate the versatility and applications of various mathematical and fuzzy methodologies across the science of decision making, management, and engineering. The work is divided into three sections: the first, composed of four chapters that use methodologies for different selection problems; the second, composed of three chapters that present methodologies for information analysis; and the third, composed of three chapters that integrate new formulations and extensions of methodologies. The book starts with a chapter that covers the fundamentals used, including the different methods and formulations applied.

In the first section, which analyzes different selection problems, the authors present methodologies to obtain a ranking based on the selected criteria and, in this way, identify the best option. Chapter 2 is an application in the financial sector, specifically on how to select the best Exchange-Traded Fund (ETF) from a set of options. The methodology used is based on the Multi-Criteria Hierarchical Process (MCHP) in conjunction with the Elimination and Choice Translating Reality (ELECTRE) III. Within the results, 20 different ETFs are analyzed, and a selection ranking is presented based on an analysis of 7 market risk indicators and 5 financial indicators. Chapter 3 presents an analysis to identify the critical factors within an Information Technology Governance (ITG) process. The methodology used was the Fuzzy Technique for Order Preference by Similarity to Ideal Solution (Fuzzy TOPSIS), and 68 critical factors were analyzed, classified into 7 key areas, resulting in the identification of 6 strategic components for a good ITG.

Continuing with the chapters related to the use of selection techniques, Chapter 4 presents the use of the Ordered Weighted Average (OWA) operator and interval numbers for the analysis and generation of ranking on social entrepreneurship based on the values of the States of Mexico, and 13 criteria are analyzed for each of the 32 states. Finally, Chapter 5 analyzes supplier selection using the TODIM (Portuguese acronym for Interactive Multi-Criteria Decision Making), Analytic Hierarchy Analysis (AHP), and Preference Selection Index (PSI) methodologies, with ten criteria and four suppliers evaluated.

The second section of the book presents chapters on information analysis methodologies. In Chapter 6, a survey was conducted among 11,078 students of the Upper Secondary Education system in Mexico to assess their perceptions of 25 indicators. Once analyzed, these perceptions identified 16 strategies for continuous improvement for educational centers. Chapter 7 provides a bibliometric analysis of food insecurity and non-communicable diseases, analyzing more than 3000 publications indexed in the Scopus database and identifying areas of study that are particularly relevant to investigate in this field. Finally, Chapter 8 presents an analysis of the Smart Tourism Destination, in which the existing information is unified through exploratory

DOI: 10.1201/9781003724902-1

research, analytical evaluation, theoretical analysis, and a panel of experts to design the Assessment Model for Information and Communication Technologies Indicators within tourist destinations, identifying 5 strategic axes composed of 49 indicators.

The last section of the book presents three chapters that developed new formulations and mathematical and methodological extensions. Chapter 9 presents a novel decision-support framework that integrates neutrosophic logic with blockchain technology to address these challenges in uncertain environments. In Chapter 10, two novel heterogeneous penalty aggregation functions are proposed. First, the penalty weight adequacy coefficients (P-WAC). Second, the combination of the P-WAC with Bonferroni OWA operators generates an operator called Bonferroni Penalty Adequacy Coefficient OWA (BPAC-OWA). These new formulations capture the importance, expectations, and disagreements inherent in decision-making processes. Finally, Chapter 11 introduces the notions of fuzzy βSC-compact, fuzzy βC-compact, and fuzzy β-almost compact spaces, extending the classical notion of β-compactness to the framework of fuzzy topological spaces.

As can be seen, the book presents, in a holistic way, different ways of applying and designing mathematical and fuzzy methodologies across decision making, administration, and engineering, with applications in selection, information analysis, and the generation of new formulations. Therefore, through this compilation, what is sought is that the reader can design their own uses of this type of methodology for business and professional applications or identify new research and application topics for this type of method, characterized by their flexibility and adaptability across different areas and problems.

1 Fundamentals

Fabio Blanco-Mesa, Ernesto León-Castro, and Victor Alfaro-Garcia

This chapter addresses three main aspects: (1) ranking and decision-making, (2) methods of information analysis, and (3) new mathematical and methodological formulations and extensions. We explain each of the approaches below.

1.1 RANKING AND DECISION-MAKING

The first approach examines various methodologies that enable the creation of rankings using multi-criteria methods. The methods used include the multi-criteria hierarchical process (MCHP) (Corrente et al., 2012), ELECTRE III (Figueira & Roy, 2002), the fuzzy technique for order preference by similarity to ideal solution (Fuzzy TOPSIS) by Chen and Hwang (1992), principal component analysis (PCA) (Mooi et al., 2018), ordered weighted average (OWA) (Yager, 1988), TODIM (interactive multi-criteria decision-making) (Gomes, 2009; Ren et al., 2016), and analytic hierarchy process (AHP) (Cooper et al., 2006).

The MCHP (Corrente et al., 2012) enables the importance of the expert's preferences to be ascribed to their role in establishing the model for developing alternatives and facilitating decision-making. Electre III (Figueira & Roy, 2002) is a multi-criteria decision analysis (MCDA) method that employs fuzzy outranking relations to rank alternatives. Similarly, ELECTRE III facilitates the determination of the "outranking" (superiority) of one alternative over another, based on the concordance (majority) and discordance (minority veto) principles. Likewise, the fuzzy technique for order preference by similarity to ideal solution (fuzzy TOPSIS) (Schmitz & Pape, 2020) is a multicriteria decision-making method as a tool to hierarchies the critical variables. This enables the capture of the ambiguity intrinsic to human judgement through the incorporation of fuzzy numbers. The establishment of a positive ideal solution and a negative anti-ideal solution serve as benchmarks to determine the relative closeness of each alternative. Also, PCA is combined with Fuzzy TOPSIS and get better results in selection of alternatives. PCA is a data mining technique that maximises the variance explained by each component, thereby facilitating the interpretation of the information (Mooi et al., 2018). This facilitates the identification of the components that empirically group the items of the instrument designed to assess the efficiency of the implementation, and to identify each component that contributes to the efficient implementation. Furthermore, the OWA method (Yager, 1988), facilitates the aggregation of information across the maximum and minimum values. The operator performs aggregation of inputs by initially arranging the arguments in the order of magnitude, from the largest to the smallest. This is followed by the application of weights to these ordered values. Consequently, a given w_i, w_i is associated with the i-th largest argument, not necessarily the i-th input position. Besides,

DOI: 10.1201/9781003724902-2

TODIM (Gomes, 2009; Ren et al., 2016) is a multi-decision-making approach which models the psychological behaviour of experts. This approach employs the value function to articulate the decision-makers' subjective perceptions of profits or losses through a relative comparison process between alternatives, incorporating risk attitudes. Finally, Analytic Hierarchy Analysis (AHP) (Cooper et al., 2006) is a decision-making methodology that assists in the structuring of complex problems into a hierarchy of objectives, criteria, and alternatives, thereby enabling decision-makers to make pairwise comparisons between elements, assign priorities, and weights that reflect the relative importance of each factor. The central premise of this study is the transformation of subjective decisions into a quantitative process. This facilitates the integration of expert judgements and historical data, thereby enabling the acquisition of a coherent and consistent assessment of the available options.

The objective of these models and methods is to incorporate the subjective assessments of experts to establish a hierarchical order through a logical sequence that allows the alternatives to be ranked. Furthermore, these models consider the importance of weighting factors to weigh up each alternative and establish the order between them.

1.2 METHODS OF INFORMATION ANALYSIS

The second section of the book examines various methodologies focused on information analysis across different fields of application, with the aim of generating useful evidence for understanding complex phenomena and formulating improvement strategies. The methods employed in this section are diverse and include a quantitative approach supported by structured instruments for measuring perception, bibliometric analysis (Yepes-Nuñez et al., 2021), as well as the mixed methods of Hernández Sampieri et al. (2014) combined with exploratory research, analytical evaluation, theoretical analysis, and the AHP by Saaty (2008).

Quantitative methodology employs numerical data and statistical techniques to describe, explain, and predict phenomena, thereby enabling the identification of relationships, patterns, and trends through the utilisation of structured instruments (Hernández Sampieri et al., 2014). This approach is complemented by mixed-methods research, understood as the integration of qualitative and quantitative procedures within a single research design. This enables interpretative exploration to be combined with the systematic measurement of variables (Fetters et al., 2013; Johnson et al., 2007), so that the understanding of the study is not limited solely to narrative description or numerical data. Moreover, bibliometric analysis constitutes a series of quantitative techniques for the study of the evolution, structure, and impact of scientific literature (Blanco-Mesa et al., 2019; Donthu et al., 2021; García-Orozco et al., 2022; Monge et al., 2022). Its applicability is based on the facilitation of optimised information retrieval, the recognition of the most relevant actors within a field of study, the identification of emerging trends and the assessment of the impact of research output. As asserted by Paéz et al. (2024), and as corroborated by the PRISMA guidelines (Page, 2021), the identification, selection, and screening of documents can be supported. In conclusion, the study encompasses modelling and evaluation approaches that incorporate structural and content analysis techniques.

These enable information to be classified, organised, and refined using systematic criteria, such as the AHP (Saaty, 2008). The AHP is a methodology for multi-criteria decision-making based on pairwise comparisons that transforms expert judgements into quantifiable priorities. This facilitates the ranking of criteria and indicators within complex structures. The fundamental premise of these approaches is the transformation of subjective assessments into quantifiable structures, thereby facilitating the integration of expert knowledge with empirical data to generate consistent and comparable evaluations.

The aim of these methodologies is threefold: firstly, to interpret perceptions, secondly, to examine scientific output, and thirdly, to compare conceptual models and prioritise indicators. It is asserted that this will provide a robust basis for evaluation, the understanding of complex problems and informed decision-making.

1.3 NEW MATHEMATICAL AND METHODOLOGICAL FORMULATIONS AND EXTENSIONS

The third section of the book contains chapters that expand upon traditional frameworks for analysis and decision-making in complex and uncertain contexts. These contributions bring together three lines of research: firstly, the integration of neutrosophic logic (Smarandache, 1998) and blockchain technology to support decision-making processes in uncertain environments; secondly, the formulation of new aggregation operators based on the Bonferroni mean (Bonferroni, 1950), the OWA operator (Yager, 1988) and penalty-based aggregation functions (Beliakov et al., 2011; Calvo & Beliakov, 2010); and thirdly, the extension of compactness to the field of fuzzy topological spaces (Hanafy, 2004).

Neutrosophic logic (Smarandache, 1998) is characterised by the explicit incorporation of independent components such as truth, indeterminacy, and falsity. These components represent not only imprecise information, but also indeterminate or contradictory information. Consequently, neutrosophic logic expands the descriptive capacity of classical logic and conventional fuzzy approaches. The relationship between blockchain technology and the decision-making process is subject to transformation, particularly in contexts involving multiple actors, where there is a risk of manipulation or information asymmetries (Casino et al., 2019). Other methodologies include Bonferroni averages, which are useful for modelling the interaction between aggregated arguments when the criteria are not independent of one another (Beliakov et al., 2007). The OWA operator, as proposed by Yager (1988), facilitates the integration of information by considering interactions, trade-offs, and the relative importance of criteria. Its extension through approaches incorporating penalties and heterogeneity enables the modelling of disagreement, uncertainty, and diversity of opinion in decision-making processes. In the domain of fuzzy topology, the extension of classical concepts such as compactness has resulted in the development of novel structures, including fuzzy β-compactness, which has expanded the range of analytical possibilities in spaces where membership is not binary but gradual (Chang, 1968; Zadeh, 1965). The utilisation of these methodologies enables the development of more robust models to represent complex decision-making scenarios, encompassing multiple criteria, incomplete information, and non-linear relationships. Furthermore, these methods facilitate the formulation of theorems, hierarchical relationships and counterexamples, thereby establishing a more rigorous mathematical foundation. This foundation enables the identification of conditions of stability, convergence, and consistency within decision-making systems.

The methodologies employed in this third section are of significance insofar as they serve to expand the range of tools available for tackling complex problems from a more rigorous, flexible, and innovative perspective. Furthermore, the value of these methodologies lies in the fact that they not only deepen conceptual understanding but also open up new possibilities for application and research in contexts where traditional approaches prove inadequate.

REFERENCES

Beliakov, G., Calvo, T., James, S. (2011). On Penalty-Based Aggregation Functions and Consensus. In: Herrera-Viedma, E., García-Lapresta, J.L., Kacprzyk, J., Fedrizzi, M., Nurmi, H., Zadrożny, S. (eds.), *Consensual Processes. Studies in Fuzziness and Soft Computing*, vol 267. Springer. https://doi.org/10.1007/978-3-642-20533-0_2

Beliakov, G., Pradera, A., & Calvo, T. (2007). *Aggregation Functions: A Guide for Practitioners*. https://link.springer.com/book/10.1007/978-3-540-73721-6

Blanco-Mesa, F., León-Castro, E., & Merigó, J. M. (2019). A bibliometric analysis of aggregation operators. *Applied Soft Computing Journal, 81*. https://doi.org/10.1016/j.asoc.2019.105488

Bonferroni, C. (1950). Sulle medie multiple di potenze. *Bollettino dell'Unione Matematica Italiana*, *5*(3–4), 267–270.

Calvo, T., & Beliakov, G. (2010). Aggregation functions based on penalties. *Fuzzy Sets and Systems, 161*(10), 1420–1436. https://doi.org/10.1016/J.FSS.2009.05.012

Casino, F., Dasaklis, T. K., & Patsakis, C. (2019). A systematic literature review of blockchain-based applications: Current status, classification and open issues. *Telematics and Informatics, 36*, 55–81. https://doi.org/10.1016/j.tele.2018.11.006

Chang, C. L. (1968). Fuzzy topological spaces. *Journal of Mathematical Analysis and Applications, 24*(1), 182–190. https://doi.org/10.1016/0022-247X(68)90057-7

Chen, S.J., & Hwang, C.L. (1992). *Fuzzy Multiple Attribute Decision Making Methods and Applications*. Springer-Verlag.

Cooper, W. W., Seiford, L. M., & Tone, K. (2006). *Introduction to Data Envelopment Analysis and Its Uses: With DEA-Solver Software and References*. Springer Science & Business Media.

Corrente, S., Greco, S., & Słowiński, R. (2012). Multiple criteria hierarchy process in robust ordinal regression. *Decision Support Systems, 53*(3), 660–674.

Donthu, N., Kumar, S., Mukherjee, D., Pandey, N., & Lim, W. M. (2021). How to conduct a bibliometric analysis: An overview and guidelines. *Journal of Business Research, 133*, 285–296. https://doi.org/10.1016/j.jbusres.2021.04.070

Fetters, M. D., Curry, L. A., & Creswell, J. W. (2013). Achieving integration in mixed methods designs-principles and practices. *Health Services Research, 48*(6 Pt 2), 2134–2156. https://doi.org/10.1111/1475-6773.12117

Figueira, J., & Roy, B. (2002). Determining the weights of criteria in the ELECTRE type methods with a revised Simos' procedure. *European Journal of Operational Research* 139, 317–326.

García-Orozco, D., Alfaro-García, V. G., Merigó, J. M., Espitia Moreno, I. C., & Gómez Monge, R. (2022). An overview of the most influential journals in fuzzy systems research. *Expert Systems with Applications, 200*, 117090. https://doi.org/10.1016/j.eswa.2022.117090

Gomes, L. F. A. M. (2009). An application of the TODIM method to the multicriteria rental evaluation of residential properties. *European Journal of Operational Research, 193*(1), 204–211. https://doi.org/10.1016/j.ejor.2007.10.046.

Hanafy, I. (2004). Fuzzy β-compactness and fuzzy β-closed spaces. *Turkish Journal of Mathematics, 28*(3), 281–294.

Hernández Sampieri, R., Fernández Collado, C., & Baptista Lucio, P. (2014). *Metodología de la investigación.* McGraw Hill España. https://dialnet.unirioja.es/servlet/libro?codigo=775008

Johnson, R., Onwuegbuzie, A., & Turner, L. (2007). Toward a definition of mixed methods research. *Journal of Mixed Methods Research, 1,* 112–133. https://doi.org/10.1177/1558689806298224

Monge, R. G., Alfaro-García, V. G., Espitia-Moreno, I. C., García-Orozco, D., & de Vivar Mercadillo, M. R. R. (2022). Environmental sustainability: A 10-year bibliometric analysis of the developments in management, business. *Finance and Economics* (pp. 309–337). https://doi.org/10.1007/978-3-030-96150-3_18

Mooi, E. A., Sarstedt, M., & Mooi-Reci, I. (2018). *Principal Component and Factor Analysis.* Springer.

Paéz, L., Suarez-Gomez, D., Hernández-Moreno, L., & Blanco-Mesa, F. (2024). Intención emprendedora en estudiantes Universitarios en Colombia: Un estudio bibliométrico. Inquietud Empresarial, *24*(2), 1–42. https://doi.org/10.19053/uptc.01211048.18081

Page, M. J. (2021). The PRISMA 2020 statement: An updated guideline for reporting systematic reviews. *Research Methods & Reporting, 372*(71). https://doi.org/10.1136/bmj.n71

Ren, P., Xu, Z., & Gou, y. X. (2016). Pythagorean fuzzy TODIM approach to multi-criteria decision making. *Applied Soft Computing, 42,* 246–59. https://doi.org/10.1016/j.asoc.2015.12.020.

Saaty, T. L. (2008). Decision making with the analytic hierarchy process. *International Journal of Services Sciences, 1*(1), 83–98. https://doi.org/10.1504/IJSSCI.2008.017590

Schmitz, C., & Pape, S. (2020). LiSRA: Lightweight security risk assessment for decision support in information security. *Computers & Security, 90,* 101656.

Smarandache, F. (1998). *Neutrosophy: Neutrosophic Probability, Set, and Logic: Analytic Synthesis & Synthetic Analysis.* American Research Press. https://philpapers.org/rec/SMANNP

Yager, R. R. (1988). On ordered weighted averaging aggregation operators in multicriteria decisionmaking. *IEEE Transactions on Systems, Man, and Cybernetics, 18*(1), 183–190. https://doi.org/10.1109/21.87068

Yepes-Nuñez, J. J., Urrútia, G., Romero-García, M., & Alonso-Fernández, S. (2021). The PRISMA 2020 statement: An updated guideline for reporting systematic reviews Declaración PRISMA 2020: Una guía actualizada para la publicación de revisiones sistemáticas. *Revista espanola de cardiologia, 74*(9), 790–799.

Zadeh, L. A. (1965). Fuzzy sets. *Information and Control, 8*(3), 338–353. https://doi.org/10.1016/S0019-9958(65)90241-X

2 Hierarchical Ranking of the Selection of International Exchange-Traded Funds

Manuel Muñoz Palma,
Pavel Anselmo Álvarez Carrillo,
Eva Luz Miranda Espinoza,
María del Refugio Bernal Agramón,
Ernesto León-Castro, and Pavel López Parra

2.1 INTRODUCTION

Currently, international financial markets offer a wide variety of investment instruments. Depending on an investor's profile and risk tolerance, various strategies can be employed to construct investment portfolios that aim to maximize returns and diversify risk (Arriaga et al., 2019). Since their first issuance on the New York Stock Exchange in 1993, Exchange-Traded Funds (ETFs) have experienced exponential growth across stock exchanges. These instruments, essentially mutual funds linked to underlying assets, are now widely used in international markets. This growth has accelerated over the past decade, particularly in the world's leading and most developed stock exchanges. This expansion is primarily driven by increased market participation and the proliferation of investment funds launched by major global brokerage firms. Consistent with this trend, the past 20 years have seen a significant rise in both the number of mutual funds involved in private markets and the total dollar value of their investments in private firms.

As previously established, we document a dramatic increase over the past two decades in both the number of mutual funds participating in private markets and the dollar value of their investments in private firms Kwon et al. (2020). The study of ETFs has garnered significant interest because, despite being a relatively nascent investment vehicle, ETFs are considered a pivotal innovation in financial markets, demonstrating robust growth. ETFs are among the most influential financial innovations currently transforming the investment fund market across numerous countries. Due to their structural similarities, ETFs can serve as substitutes for traditional mutual funds (Marszk et al., 2017). The primary function of ETFs is to replicate a specific market or aim to deliver comparable performance.

DOI: 10.1201/9781003724902-3

Their inherent characteristics, such as a low-cost structure and high liquidity, make them appealing to both small individual investors and large institutions alike (Liebi, 2020).

This study aims to analyze various ETF assets to evaluate their key financial indicators for investment portfolio selection. The portfolio comprises 20 shares from major stock ex-changes across diverse regions, including North America, South America, Europe, and Asia. This evaluation is based on the main financial ratios of the risk market. The selection is performed using the multiple criteria hierarchical process (MCHP) and ELEC-TRE III method. The MCHP method, which extends ELECTRE III to a hierarchy of interactive criteria, helps decision-makers (DMs) assign weights to the primary financial indicators.

The MCHP method allows the evaluation and classification of significant ETF assets for investment across the world's leading stock exchanges. This process generates a ranking of top ETF assets for investment, sound for major global financial brokerages, key DMs, stock exchange directors, and general investors.

The chapter is structured as follows: Section 2.2 presents the literature review, and Section 2.3 details the methodology. The selection of ETF assets using the multicriteria hierarchical process (MHPs) for portfolio integration is presented in Section 2.4. Finally, Section 2.5 offers conclusions and recommendations.

2.2 LITERATURE REVIEW

This section identifies the primary issues associated with ETFs and the methodologies employed to address them, drawing on a review of recent literature.

The development of ETFs in emerging economies is a significant area of study. For in-stance, Lechman and Marszk (2015) analyzed the relationship between information and communications technologies (ICT) and financial innovation development in emerging economies. Their methodological framework included descriptive statistics and logistic growth models to estimate patterns of ETF market development. Their findings indicate a robust positive growth in ICT and ETFs in Japan, Mexico, the United States, and South Korea. Conversely, in Brazil, the relationship between ICT and ETF market development, though still positive, is weaker.

Ruiz and Perales (2022) explored the potential of the MIST countries (Mexico, Indonesia, South Korea, and Turkey) to become leading emerging market players by studying their ETF markets. This research utilized financial time series analysis, identifying outliers, ap-plying volatility models, and performing time series decomposition.

Marszk et al. (2017) provided insights into the development of the ETF market in Mexico from 2002 to 2012. Using descriptive statistics and logistic growth models, they found strong development in asset management and market share for this financial innovation. Furthermore, their analysis of ETF market development trends predicted greater market share growth in the medium term (the next 3 to 5 years), albeit with a low probability of exceeding 20–30%.

Another research avenue explores whether ETF investments are more profitable than traditional strategies. Arriaga et al. (2019) posited that an investment strategy incorporating ETFs is more profitable than a traditional approach due to the nature of these financial vehicles and their low commissions. Their analysis covered Germany, the United Kingdom, the United States, and Mexico. Similarly, Wu et al. (2021) found that the general strategy for individual investors in the Chinese stock market involves choosing among multiple passive vehicles that track an index, with their results indicating that ETFs outperform index funds.

Gleason et al. (2004) analyzed intraday data to investigate whether investors exhibit herding behavior during periods of extreme market movements, specifically, Exchange-Traded Funds (ETFs). The authors identify extreme up-market and down-market periods, and the other incorporate a nonlinear term in a regression specification to detect potential herding in nine sector ETFs traded on the US Stock Exchange. The findings confirm the hypothesis that investors do not behave as herds during the periods studied.

Further research highlights various aspects of ETF performance and their role in financial markets. Malhotra (2024) examined the net monthly returns of real estate ETFs using di-verse performance evaluation models and various market situations. The results indicated that these ETFs generated positive alphas and outperformed benchmark indices in absolute returns. The prospects for global sustainable investing appear favorable, spurred by ongoing environmental initiatives at the supranational level in both developed and leading developing countries (Khmyz, 2024). In response to the European debt crisis and its aftermath, there has been an increased use of Gold ETFs and government-issued Gold Bonds as risk management tools (Khindri, 2024). Additionally, Wang et al. (2022) found that when clean energy ETFs or natural gas are utilized, the predictive information de-rived from global economic conditions surpasses that of traditional uncertainty indices.

Regarding the discussion around introducing a Bitcoin ETF, Blazsek and Bowen (2024) demonstrated that score-driven portfolios outperform both classical volatility model-based portfolios and equity portfolios. Furthermore, an evaluation using global ETF data by Wu et al. (2021) indicated that their proposed Generative Ensemble Model (GEM) with joint cost outperformed the equally weighted portfolio and the ordinary deep learning model by 33.7% and 30.1%, respectively. Fisch et al. (2020) argue that institutional passive investor ETFs represent the most significant development in modern capital markets, directing trillions of dollars in capital flows and increasingly owning substantial portions of the US corporate sector. Sezer and Ozbayoglu (2018) evaluated the effect of volatility on ETF prices, finding a negative correlation between stock prices in the market and increased volatility.

In recent years, the ETF market has experienced exponential growth, soaring from $93 billion in 2020 to $306 billion in 2022 (Liu, 2023). Figure 2.1 illustrates the investment evolution of ETFs on the New York Stock Exchange. This reflects a considerable increase in investment, primarily driven by brokerage firms. In this sense, the advantages of investing in ETFs allow several benefits for investors, which are highlighted in Table 2.1.

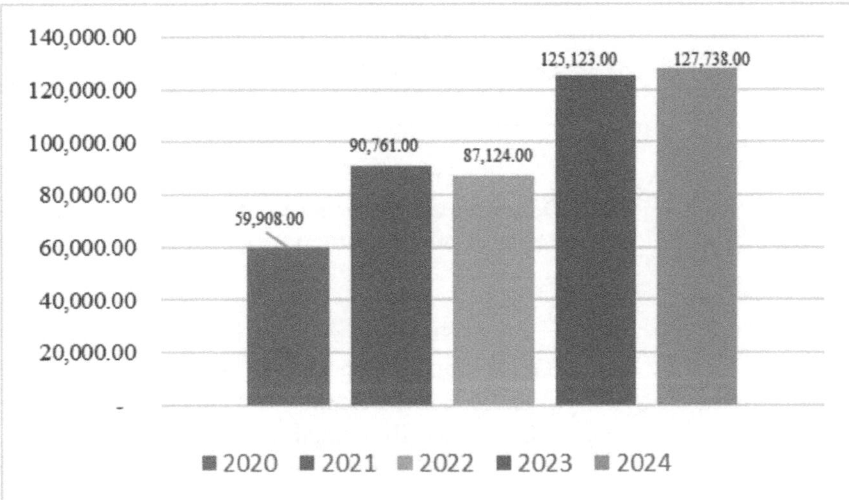

FIGURE 2.1 Investment evolution of ETFs on the New York Stock Exchange. (Own elaboration based on nyse.com.)

TABLE 2.1
Main Advantages of ETFs for Investors

Benefits	Description
Accessibility	ETFs are relatively affordable, with low management fees.
Diversification	ETFs allow investing in a wide range of assets, which helps reduce risk.
Liquidity	ETFs can be easily bought and sold on the stock exchange every day from Monday to Friday and during the active trading day.

2.3 METHODOLOGY

Multicriteria analysis involves comparing different alternatives based on selected criteria. In this process, the expert or DM constructs a set of alternatives informed by their preferences and experiences in the field under investigation.

2.3.1 Multicriteria Hierarchical Process

The importance of the expert's preferences lies in their role in establishing the model for developing alternatives and facilitating decision-making. MCHP is applied using various criteria to evaluate key ETF assets for integration into an investment portfolio. These criteria include financial ratios (C1) and market risk (C2). Additionally, 12 sub-indexes representing the behavior performance of different companies that issue ETF assets from major stock exchanges in 2024 are considered.

Figure 2.2 illustrates the five stages involved in preparing and obtaining the ranking for this research. Stage 1 involves evaluating the leading financial indicators of

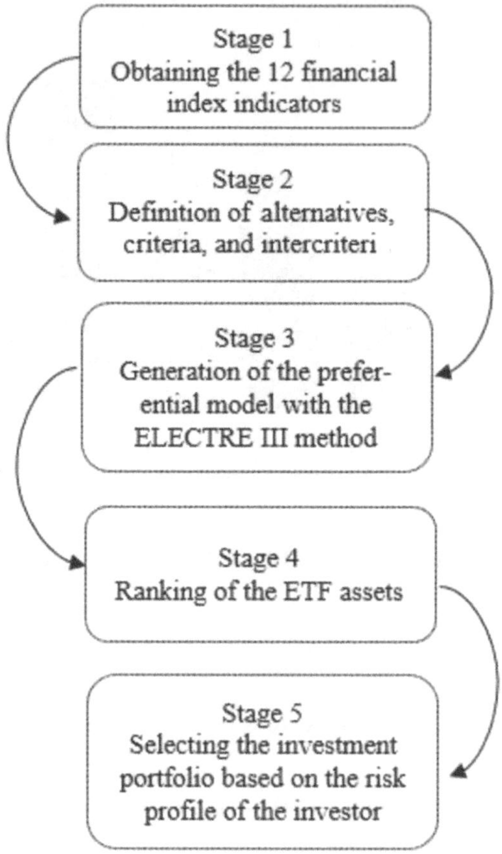

FIGURE 2.2 Research model for selecting the main ETF assets.

company performance on stock exchanges, with data sourced from Investing.com. In Stage 2, the 12 performance criteria for the financial indicators of ETF-issuing companies are defined, along with the parameters for the ELECTRE III multicriteria method. Stage 3 applies the ELECTRE III method to generate a preference model. Stage 4 focuses on exploiting this preferential model, employing a distillation process that yields a ranking of the main ETF assets for investment portfolio integration. Finally, in Stage 5, the primary assets that will constitute the ETF investment portfolio are determined based on the generated ranking and performance analysis.

The MCHP is rooted in MCDA (Corrente et al., 2012), particularly for practical applications involving a hierarchical structure of criteria. This hierarchical approach is well-suited to the inherently hierarchical nature of many decision problems. Consequently, the MCHP's relevance lies in its ability to generate a hierarchical structure, facilitating the creation of subgroups from a large number of criteria. Within the MCDA process, developing a clear set of alternatives $A = \{a_1, a_2, \ldots, a_m\}$ and a coherent family of criteria $G = \{g_1, g_2, \ldots, g_m\}$ is crucial for effective decision-making. This approach is beneficial for developing strategies of selecting and integrating investment portfolios of ETF assets and for ranking assets based on their financial ratio performance. Indeed, practical applications often necessitate such a hierarchical structure (Corrente et al., 2012).

A multicriteria analysis method, in its classical approach, typically analyzes stocks at a single level, evaluating all criteria simultaneously. While this can identify the best and worst performing ETF assets based on financial ratios from leading companies, it fails to illuminate how specific sub-criteria (index subgroups) (e.g., market factors and financial ratios) interact at different levels to influence the selection of ETF assets. Therefore, a distinct method like the MCHP, which allows for evaluating stocks by subsets of criteria at varying levels, would be valuable for solving the problem of selecting ETF assets for investment portfolio integration.

2.4 SELECTION OF ETF ASSETS WITH THE MULTICRITERIA HIERARCHICAL PROCESS

This section examines the financial ratios of 20 companies that issue ETFs in the 2024 period. Table 2.2 shows the 20 main assets. We have compiled statistical reports for the data, reflecting their performance on stock exchanges, which are categorized into two dimensions: market risk (g1) and financial indicators (g2). These dimensions are used to evaluate ETF assets listed in the indexes, primarily those on

TABLE 2.2
ETF Assets as Decision Alternatives

Label	ETFs Assets
A1	Alerian MLP ETF (AMLP)
A2	Invesco QQQ Trust (QQQ)
A3	Brazil ETF (EWZ)
A4	Eurozone ETF (EZU)
A5	WisdomTree India Earnings Fund (EPI)
A6	WisdomTree Europe Hedged Equity Fund (HEDJ)
A7	SPDR S&P Regional Banking ETF (KRE
A8	iShares MSCI Japan ETF (EWJ)
A9	China Large-Cap ETF (FXI)
A10	MSCI Taiwan ETF (EWT)
A11	First Trust Latin America Alpha DEX Fund
A12	MSCI South Africa ETF (EZA)
A13	Franklin FTSE Latin America ETF (FLLA)
A14	First Trust Nasdaq Semiconductor ETF (FTXL)
A15	Global X MSCI Colombia ETF (GXG)
A16	VanEck Indonesia Index ETF (IDX)
A17	MSCI Turkey ETF (TUR)
A18	MSCI Mexico ETF (EWW)
A19	First Trust Brazil Alpha DEX Fund (FBZ)
A20	Semiconductor ETF (SOXX)

TABLE 2.3
Macrocriteria and Elementary Criteria for Selecting ETF Assets

Macrocriteria	Index	Elementary Criteria	Weights
Market risk (g1)	g1,1	Alfa	0.065
	g1,2	BETA	0.065
	g1,3	Mean annual return	0.03
	g1,4	R-squared	0.02
	g1,5	Standard deviation	0.07
	g1,6	Sharpe ratio	0.07
	g1,7	Treynor ratio	0.08
Financial indicators (g2)	g2,8	Variation in one year	0.09
	g2,9	Return on investment	0.13
	g2,10	Total assets	0.14
	g2,11	Market capitalization	0.12
	g2,12	Dividends	0.12

the New York Stock Exchange and other major exchanges worldwide, encompassing various countries across America, Eu-rope, Asia, and Africa. Each dimension comprises a subgroup of different indicators (elementary criteria); in total, 12 indicators are used to evaluate the shares of the leading ETF-issuing companies on stock exchanges. This shows in Table 2.2.

This chapter adopts a novel approach, employing the MCHP to analyze market indices. This method uniquely considers the interaction of criteria subgroups at different levels within a hierarchy, enabling the ordering of ETF assets. Table 2.3 also details the macrocriteria, elementary criteria, and their corresponding weights for the stock selection problem.

The weights definition is carried out with the deck card procedure first proposed by Simos (1990) and then improved with the new version Simos' revised procedure by Figueira and Roy (2002) that is known as SRF by the software acronym of Simos Roy Figueira. The procedure is carried out in two stages. First stage, the DM or the analyst need to order the financial indicator from the least important to the most important. If it is needed some criteria belong to a set of criteria. So, the DM order criteria, groups of criteria or both a combination of groups and individual criteria are ordered.

The second stage is the weights determination. The desk card procedure assigns less weights to the first criterion or the group of criteria and highest weight value to the last criterion or group of criteria. The criteria in the same group are assigned the same weight value.

The financial indicators were ordered as follows: $g1,4 \prec g1,3 \prec \{g2,8 \approx g2,9\} \prec \{g1,1 \approx g1,2 \approx g2,10 \approx g2,11\} \prec \{g1,5 \approx g2,12\} \prec g1,6 \prec g1,7$. The criteria in the same group are assigned the same criteria value as shown as follows 0.05, 0.06, {0.07, 0.07}, {0.08, 0.08, 0.08, 0.08}, {0.1, 0.1}, 0.11, 0.12.

Following the methodology outlined in Section 2.3, the MCHP is applied to select ETF assets for integration into investment portfolios of major stock exchange companies at a global level. Initially, the problem is structured as a multicriteria hierarchy, breaking it down into the macrocriteria market risk (g1) and financial indicators (g2). This acts as a sub-problem for the assets. It is illustrated in the hierarchical structure in Table 2.3. Hither, shares of companies listed on stock exchanges are organized hierarchically based on these 2 macrocriteria and 12 elementary criteria. This new hierarchical structure for asset selection brings the analysis closer to the core principles of the MCHP.

This approach evaluates each macrocriterion, enabling an analysis of the direct interaction between its immediate descending subcriteria. Simultaneously, it involves generating preferential models and rankings for each macrocriterion. This helps us understand how one ETF asset performs relative to another and, crucially, how this impacts the selection of ETF assets for investment portfolio integration tailored to an investor's profile.

2.4.1 ETF Asset Analysis with the Multicriteria Hierarchical Process

ELECTRE III method was applied to solve each subproblem g_i (macrocriterion) and to derive an integrated overall ranking (see Table 2.4). Each macrocriterion is evaluated by a subset of sub-criteria (elementary criteria at the lowest level of the hierarchy). This process yielded an integral ranking (g0) and individual (g1, g2) rankings of the 20 ETF assets from major stock exchanges. The overall ranking is a result of

TABLE 2.4
Comprehensive (g_0) and Individual (g_1, g_2) Rankings of ETF Assets

Position	g_0	g_1	g_2
1	A2	A20	A10
2	A1, A11	A14	A2, A11
3	A17	A1	A1
4	A20	A18	A15
5	A10	A19	A3, A8
6	A4	A11	A17
7	A15	A6	A13
8	A18	A4	A20
9	A3	A17	A7
10	A13	A15	A18
11	A6	A2	A5, A14
12	A8, A19	A3	A19
13	A4	A10	A6
14	A7	A12	A12
15	A12	A5	A4
16	A5	A13	A9
17	A9	A7	A16
18	A16	A9, A8	
19		A16	
20			

TABLE 2.5
Performance of Market Risk Indicators in the Top Positions of the Ranking (g_0)

Label	C_1	C_2	C_3	C_4	C_5	C_6	C_7
A1	18.99	0.63	2.11	25.00	20.90	1.06	35.82
A2	0.04	1.18	1.19	85.67	22.54	0.50	7.96
A10	3.10	1.17	0.53	67.57	23.33	0.15	0.76
A11	8.84	1.05	1.01	51.12	23.95	0.38	6.36
A17	21.60	0.45	2.06	4.46	34.71	0.63	39.60
A20	10.65	1.63	1.83	66.38	33.21	0.58	9.20

the intricate interaction among these elementary criteria, which collectively evaluate their respective macrocriteria. For the ETF asset selection problem, our analysis examined how the interaction of elementary criteria influences the macrocriteria (Level 2 of the hierarchy). Subsequently, we investigated the impact of these macrocriteria on the stock selection problem, specifically in the context of creating ETF investment portfolios (Level 1).

The top positions in the integral ranking (g0) results include Invesco QQQ Trust (A2), ranked first, followed by a tie among Alerian MLP ETF (A1) and Franklin FTSE Latin America (A11), ranked second, followed by MSCI Turkey ETF (A17), and Semiconductor ETF (A20), ranked third and fourth, respectively. These five assets represent the leading candidates for integration into an investment portfolio. Invesco QQQ (A2) notably stands out for its strong financial performance. It ranks first within the market risk macrocriterion (g1) and particularly excels in the elementary performance criteria R-squared (C4), standard deviation (C5) and Sharpe ratio (C6). Similarly, the Financial Ratios macrocriterion (g2) holds the top position across indicators Variation in one year (C8), Total assets (C10), and Dividends (C12) as shown in Tables 2.5 and 2.6, which is derived from the overall ranking presented in Table 2.4.

Before a description of the outcomes, it is worth to mention some tides on the order shown on Table 2.4. The ELECTRE-III is an outranking approach that compare alternatives in pairwise comparison, different to the full aggregation approach (e.g., additive methods) where each alternative is assigned a global value to be

TABLE 2.6
Performance of Financial Indicators in the Top Positions of the Ranking (g_0)

Label	C_8	C_9	C_{10}	C_{11}	C_{12}
A1	0.2242	0.0743	8.048	8.4600	3.5200
A2	0.3454	0.0054	43.640	248.2800	2.2900
A10	0.0790	0.1188	3.580	4.0900	5.5270
A11	0.1142	0.0388	127.86	62.0200	0.7512
A17	0.2351	0.0369	0.41953	0.2306	1.4307
A20	0.6073	0.0042	0.55822	12.3800	0.8998

evaluate (e.g., as utility value). In this sense, the outranking method compares the performance of alternatives in each criterion regarding the assessment alternative *a* is as good as *b*. The distillation procedure uses an input the fuzzy outranking matrix computing some power and weakness of the alternatives compared against the rest. Once the computation is made, the distillation method generate two ranking, the descendent and ascendent ranking. Those ranking are evaluated with the following sentences to stablish if an alternative *a* is in better (strict preferred), same (indifference) or incomparable to *b*.

- *a* is strictly preferred to *b* if *a* is better positioned than *b* in at least one of the rankings, and if *a* is at least as good as *b* in the other rank.
- *a* is indifferent to *b* if *a* and *b* are placed in the same position (belong to same group) in the two rankings.
- *a* is incomparable to *b* if *a* is better positioned than *b* in one ranking and *b* is better positioned than *a* in the other ranking.

A complete step-by-step description of the distillation method is given in Alvarez et al. (2022), and the computational tool is available on https://github.com/paac80/Simos_Revised.

A complete description step by step of the distillation method is described in Alvarez et al. (2022), and the computational tool is available on https://github.com/paac80/Simos_Revised_Procedure.

An specific weight was assigned to g1 and g2 based on the relative importance of each macrocriterios, these weights are 0.60 for g1 and 0.40 for g2. The stock market index indicators for g1 reveal the positions of the ETF assets in the following order: A20 > A14 > A1 > A18 > A19 > A11 > A6 > A4 > A17 > A15 > A2 > A3 > A10 > A12 > A5 > A13 > A7 > A9 = A8 > A16. Based on the classification ordering of ETFs, according to the performance of the macro criterion g1, of the performance of the Market Risk indicators, where investment brokerages define their portfolios according to the performance perspectives and risk profile of the investor, it is important to note that investing in ETFs represents a general investment risk. However, the ETF market is poised for significant development due to the advantages it offers investors, primarily in terms of liquidity, market trading, and low transaction costs, among other benefits.

Meanwhile, MSCI Taiwan ETF (A10) holds the first place within the Financial Ratio macrocriterion (g2). This company demonstrates superior performance in the investment return criterion (C9) with 11.88%, compared to (A2)'s 0.54%, and also excels in the dividend payment criterion (C12). These figures highlight significant differences between the ETF assets. However, regarding total assets (C10), ETF QQQ (A2) shows greater strength relative to assets with 43.64, while ETF (A10) has 3.58, indicating (A2) is approximately 12 times larger. For investors, this represents a crucial indicator for assessing coverage capacity, ensuring the ability to meet payment commitments, and reducing payment risk (see Table 2.6).

Analyzing the assets at the lowest ranks of the obtained ELECTRE III ranking, it is notable that ETF investment stands as a primary option for integrating investment portfolios across most of the world's leading brokerages. BlackRock, a prominent global investment fund brokerage, has particularly distinguished itself through its substantial participation in ETF portfolios over the last three decades. Demonstrates that ETFs are an investment strategy capable of reducing risk and maximizing return potential compared to direct stock purchases, ultimately aiming for greater diversification in asset investment.

The global accessibility of the ETF market, and specifically, for this research, the availability of global ETFs from emerging countries, coupled with their attractive liquidity (ease of converting an investment to cash), further enhances their appeal. Another key feature for investors is the flexibility of these instruments' operability; that is, the ability to change short-term positions allows for seamless market entry and exit, while long-term positions prove highly profitable. Furthermore, the advantage of paying significantly lower commissions, often representing the lowest rates in the market, is also notable. For example, iShares Core ETFs, on average, incur only one-tenth the expense of most equity investment funds. However, like all investments in financial instruments generally bought and sold on global stock markets, ETFs carry inherent risks. Specifically, global ETF assets are exposed to various risks, including economic, currency, and liquidity risks.

As illustrated in Table 2.4, the bottom five positions in the overall ETF ranking (g0) are held by assets A7 > A12 > A5 > A9 > A16. Specifically, the VanEck Indonesia Index ETF IDX (A16) occupies the 18th position, and China Large-Cap ETF FXI (A9) is ranked 17th. Generally, global emerging market ETF assets tend to occupy the lower ranks, with those at the very bottom being geographically located in Asia.

Most global ETFs are exposed to various risks, including volatility, country-specific factors, sector exposure, liquidity, exchange rate fluctuations, and broader economic concerns, among others. In Table 2.7, we analyze the case of the China Large-Cap ETF FXI (A9), which primarily comprises assets from the 50 largest companies listed on the Hong Kong Stock Market. In this sense, market risk macrocriteria reveal a high volatility of 32.7 (C3) and a negative return of –1.18. Consequently, this higher volatility, indicating greater risk, results in the China Large-Cap ETF FXI asset having lower average profitability. Although its Beta (C2) is 1.02, this indicates it tracks the market. The interpretation of Beta coefficients requires understanding their macroeconomic essence and the factors that determine them (Lukina, 2023). This is particularly crucial for the Chinese market, given the limited reliable information due to state control over economic activities and the significant regulatory and political risks that impact the markets. Furthermore, China's GDP growth

TABLE 2.7
Performance of Market Risk in the Bottom Positions of the Ranking (g_0)

Label	C_1	C_2	C_3	C_4	C_5	C_6	C_7
A7	–9.41	0.92	–0.15	24.84	30.34	–0.15	–10.30
A12	–4.70	1.24	–0.12	62.42	25.68	–0.17	–6.02
A5	–11.36	0.54	1.20	38.45	14.34	0.81	20.90
A9	–17.39	1.06	–1.18	27.95	32.70	–0.52	–19.33
A16	4.70	0.45	–0.14	26.13	14.08	–0.32	12.44

is currently slowing to 4.3%, partly due to the effects of US government global tariff policies and "trade war" policies, according to data published by the World Bank.

According to World Bank projections, growth in East Asia and the Pacific is expected to be 4.8% in 2024 and 4.4% in 2025. Given that markets in this region are highly concentrated in the financial and technology sectors, both the VanEck Indonesia Index EF (A16), and China Large-Cap ETF (A9) show negative results when evaluated against macrocriteria performance using the Sharpe ratio (C6). Hence, its expected rate of return is lower than the risk-free rate, indicating suboptimal management. This implies that investors taking a risk on these ETFs might find it unwarranted when compared to other investment alternatives. Utilizing a Multicriteria Decision-Making (MCDM) method can yield results that facilitate improved decision-making in portfolio integration by assessing market risk based on financial indicators to evaluate risk, return, and Sharpe ratios (Silva et al., 2023).

Regarding the performance of the financial ratios macrocriterion (g2) for the lowest-ranked positions, the financial indicator performance of the VanEck Indonesia Index ETF IDX (A16) and China Large-Cap ETF (A9). These assets occupy the last positions in the overall ranking (g0), as well as in the market risk (gl) and financial indicators (g2) macrocriteria. In Table 2.8, we analyze the performance within the Financial Indicators macrocriterion (g2), the China Large-Cap ETF (A9) asset exhibited a negative average annual variation of -8.19%, despite its underlying criteria (C10, C11, and C12) being positive. Based on our analysis, ETF (A9) represents 6.9% of total invested assets, boasts a market capitalization (C11) of 4.61, and has a dividend pay-out (C12) of 0.9108. In stark contrast to ETFs at the bottom of the overall ranking (g0) and within macrocriterion (gl) and (g2), the Indonesian asset, the VanEck Indonesia Index ETF, generally showed a negative average return (C8) of –15.27%. This performance accounts for 8.0% This performance accounts for an 8.0% contribution to the ranking weight using the ELECTRE III method.

Ultimately, ETFs generally serve as a viable investment option for funds and brokerages, aligning with diverse investor profiles due to their inherent advantages. These include, but are not limited to, liquidity, operational ease, and low fees. However, it is crucial to acknowledge the associated risks in global markets and various geographic regions, such as the impact of government policies and exposure to foreign exchange and economic risks, among others.

TABLE 2.8
Performance of Financial Indicators in the Bottom Positions of the Ranking (g_0)

Label	C_8	C_9	C_{10}	C_{11}	C_{12}
A7	14.86%	3.12%	2.3300	2.340	1.5141
A12	–6.64%	3.04%	0.5040	0.2536	1.8668
A5	38.06%	0.53%	1.4320	2.9900	0.2400
A9	–8.19%	3.58%	6.7900	4.6100	0.9108
A16	–15.27%	3.89%	0.8823	0.2701	0.6176

TABLE 2.9
Value Parameters Modifications

Macrocriteria	Index	Elementary Criteria	Weights	Weight of Run 1	Weight of Run 2
Market risk (g1)	g1,4	R-squared	0.02		0.025
	g1,5	Standard deviation	0.07		0.065
Financial indicators (g2)	g2,8	Variation in one year	0.09	0.10	
	g2,10	Total assets	0.14	0.13	

2.4.2 Sensibility Analysis

A sensibility analysis was carried out for verification of the selected weights in the test. Two runs were performed with different value weights. In the first run two elementary criteria weights were modifies, the Variation in one year (g2,8) and Total assets (g2,10) from the financial indicators macrocriterion. Table 2.9, run 1 column shows the new values defined. In the second run two elementary criteria weights were modifies, the R-squared (g1,4) and the Standard Deviation (g1,5) from the market risk macrocriterion. Run 2 from Table 2.9 shows the selected values.

The first run generated a ranking with no changes on position of alternatives. Column Run 1 from Table 2.10 shows the generated ranking. The second run considering

TABLE 2.10
Outcomes of Sensibility Analysis

Position	Original	Run 1	Run 2
1	A2	A2	A2
2	A1, A11	A1, A11	A1, A11
3	A17	A17	A17
4	A20	A20	A20
5	A10	A10	A10
6	A4	A4	A14
7	A15	A15	A15
8	A18	A18	A18
9	A3	A3	A3
10	A13	A13	A13
11	A6	A6	A6, A19
12	A8, A19	A8, A19	A4
13	A4	A4	A7, A8
14	A7	A7	A12
15	A12	A12	A5
16	A5	A5	A9
17	A9	A9	A16
18	A16	A16	-
19	-	-	-
20	-	-	-

changes on the market risk macrocriterion shows some smooth changes on alternatives A8 and A19. Those alternatives were tied in the original ranking. In second run of the sensibility analysis, they are tied with others alternatives. In second run A8 is tied with A7, A19 is tied with A6. The differences on the current ranking are with two alternatives, one position of distance for each alternative.

2.5 CONCLUSIONS

This chapter examines the performance of companies that issue Exchange Traded Fund (ETF) assets listed on major stock exchanges. It evaluates the factors affecting ETF asset performance using two macrocriteria and 12 elementary criteria. From a methodological perspective, the MCHP was employed to analyze the performance of key stocks listed on worldwide exchanges. The analysis process yields a preferential model, which generates an ordered list for selecting the primary ETF assets suitable for integration into an investment portfolio.

Finally, the main financial ratios of leading global ETF-issuing companies were evaluated. To address complex real-world classification problems, the ELECTRE III method was extended to a hierarchical structure of criteria. This extension facilitates a more detailed analysis of preference relationships within a set of alternatives at different levels, which is particularly relevant for the ETF asset selection problem.

REFERENCES

Alvarez, P.A., Valdez, C., & Dutta, B. Analysis of the innovation capacity of Mexican regions with the multiple criteria hierarchy process Socio-Economic Planning Sciences, 84 (2022).

Arriaga, R., Castro, J.E., & Sosa, M. Analysis of investment strategies of international diversification: Traditional portfolios vs ETFs. Analisis Economico, 41–61 (2019).

Blazsek, S., & Bowen, R. Score-driven cryptocurrency and equity portfolios. Applied Economics, *56*(18) (2024).

Corrente, S., Greco, S., & Słowiński, R. Multiple criteria hierarchy process in robust ordinal regression. Decision Support Systems, 53 (3), 660–674 (2012).

Figueira, J. & Roy, B. Determining the weights of criteria in the ELECTRE type methods with a revised Simos' procedure. European Journal of Operational Research, 139, 317–326 (2002).

Fisch, J., Hamdani, A., & Solomon, S. D. The New Titans of Wall Street: A theoretical framework for passive investors. University of Pennsylvania Law Review, 168(1) (2020).

Gleason, K.C., Mathur, I., & Peterson, M. Analysis of intraday herding behavior among the sector ETFs. Journal of Empirical Finance, 11 (5), 681–694 (2004). https://doi.org/10.1016/j.jempfin.2003.06.003

Kang, Y. S., & Seok, J. Portfolio management via two-stage deep learning with a joint cost. Expert Systems with Applications, 143 (2020). https://doi.org/10.1016/j.eswa.2019.113041

Khindri, R. Testing the hedging effectiveness of the gold derivative market in India. Journal of Commerce and Accounting Research, 13(1) (2024). https://doi.org/10.21863/jcar/2024.13.1.004

Khmyz, O. Sustainable investments and divestments by global institutional investors. Financial Journal, 16(1) (2024). https://doi.org/10.31107/2075-1990-2024-1-96-108

Kwon, S., Lowry, M., & Qian, Y. Mutual fund investments in private firms. Journal of Financial Economics, 136 (2), 407–443 (2020). https://doi.org/10.1016/J.JFINECO.2019.10.003

Lechman, E., & Marszk, A. ICT technologies and financial innovations: The case of exchange-traded funds in Brazil, Japan, Mexico, South Korea, and the United States. Technological Forecasting and Social Change, 99 (2015). https://doi.org/10.1016/j.techfore.2015.01.006

Liebi, L. The effect of ETFs on financial markets: A literature review. Financial Markets and Portfolio Management, 34 (2), 165–178 (2020). https://doi.org/10.1007/s11408-020-00349-1

Lukina, Y, Investment Attractiveness Evaluation of High Growth Companies in Times of Economic Crisis (February 20, 2023). Available at SSRN: https://ssrn.com/abstract=4373264 or http://dx.doi.org/10.2139/ssrn.4373264

Liu, S. Do investors and managers of active ETFs react to social media activities? Finance Research Letters, 51 (2023). https://doi.org/10.1016/j.frl.2022.103454

Malhotra, D. Evaluating the performance of real estate exchange-traded funds. Journal of Risk and Financial Management, 17(1) (2024). https://doi.org/10.3390/jrfm17010007

Marszk, A., Lechman, E., & Kaur, H. Financial markets diffusion patterns. The case of Mexican investment funds. Equilibrium. Quarterly Journal of Economics and Economic Policy, 12 (1) (2017). https://doi.org/10.24136/eq.v12i1.5

Ruiz, C., & Perales, G. Financial emerging markets revisited. In Data Analytics Applications in Emerging Markets (2022). https://doi.org/10.1007/978-981-19-4695-0_4

Sezer, O. B., & Ozbayoglu, A. Algorithmic financial trading with deep convolutional neural networks: Time series to image conversion approach. Applied Soft Computing Journal, 70 (2018). https://doi.org/10.1016/j.asoc.2018.04.024

Silva, N. F., Dos Santos, M., Gomes, C. F. S., & De Andrade, L. P. (2023). An integrated CRITIC and grey relational analysis approach for investment portfolio selection. Decision Analytics Journal, 8, 100285.

Simos, J. Evaluer l'impact sur l'environnement: Une approche originale par l'analyse multicritere et la negociation. Presses Polytechniques et Universitaires Romandes, Lausanne (1990).

Wang, J., Ma, F., Bouri, E., & Zhong, J. Volatility of clean energy and natural gas, uncertainty indices, and global economic conditions. Energy Economics, 108 (2022). https://doi.org/10.1016/j.eneco.2022.105904

Wu, C., Xiong, X., & Gao, Y. Performance comparisons between ETFs and traditional index funds: Evidence from China. Finance Research Letters, 40 (2021). https://doi.org/10.1016/j.frl.2020.101740

3 Decision Model for the Implementation of IT Governance

An Approach Based on Fuzzy TOPSIS and Principal Component Analysis

Patricia Cristina Parroquin-Amaya, Iván Juan Carlos Pérez-Olguín, Luis Carlos Méndez-González, and Victor Manuel Morales Rocha

3.1 INTRODUCTION

In the last decades, information technologies (IT) have consolidated as an essential component in public and private organizations, generating a critical dependence on their efficient operation. Institutional productivity, profitability, and competitiveness are increasingly determined by the strategic use of IT, which requires robust and well-defined governance structures (Marulanda-Echeverry et al., 2017; Vejseli & Rossmann, 2017). In this context, ITG becomes an indispensable mechanism for aligning the use of technologies with the strategic objectives of the organization, maximizing their value and adequately managing the associated risks (Durán & Ignacio, 2015).

Several studies agree that the application of standards or frameworks for IT governance (ITG) cannot be assumed uniformly, since their effectiveness is strongly influenced by contextual factors specific to each organization, such as its size, level of technological maturity and particular operating environment (Azhar et al., 2021). In this context, the identification of *critical success factors* (CSF) is essential, as it allows focusing implementation and evaluation efforts on those elements that potentially generate a greater impact on organizational results.

Given the impact that the ITG has on organizational performance, it is a priority to identify and prioritize the critical factors that affect its effective implementation. Understanding which variables should be managed with greater emphasis and which are the main components that make up the ITG allows to properly guide decision making, optimize the allocation of resources and strengthen the alignment between IT and business.

DOI: 10.1201/9781003724902-4

This chapter presents a multicriteria decision model developed from the methodological integration of the Fuzzy TOPSIS method and the principal component analysis (PCA). The proposed methodology allows ranking a set of 68 critical factors for the efficient implementation of ITG, and grouping them into six key areas, determined by PCA. The resulting model offers IT managers a structured and prioritized tool that facilitates the management of the most relevant dimensions of governance, contributing to its effectiveness and alignment with institutional goals.

3.2 LITERATURE REVIEW

ITG has been widely recognized as a determining factor to ensure that technological investments are aligned with the strategic objectives of organizations (Weill & Ross, 2004). In this sense, multiple studies have identified critical factors that affect the effective implementation of ITG, which may vary according to the methodological approach and the application sector (Parroquín-Amaya & Pérez-Olguín, 2025).

For the development of this model, a systematic literature review (SLR) was conducted to identify and synthesize the critical factors mentioned recurrently in previous research. The review was based on recognized academic databases and was structured according to the methodological stages defined by the PRISMA method (Page et al., 2021; Rethlefsen et al., 2021), establishing inclusion and exclusion criteria and evaluating the quality of the selected studies.

A total of 68 critical factors were identified, which were classified within the relational mechanisms structure, processes and relational mechanisms (Van Grembergen et al., 2004). This literature review analyzes studies of critical factors of ITG conducted in the last two decades. The 68 identified factors can be found in Appendix A.

The classification of the factors into six dimensions was subsequently complemented by PCA, with the aim of reducing the dimensionality of the assessment instrument and improving its internal structure (Jolliffe & Cadima, 2016). Thus, an empirical grouping of the factors was obtained, aimed at facilitating their management and interpretation by IT managers.

This literature review shows that, although there are numerous proposals for ITG frameworks, there is still a need for models that integrate multicriteria approaches and data reduction techniques to facilitate decision making (Alshahrani et al., 2022; Parroquín-Amaya & Pérez-Olguín, 2025). Therefore, this chapter proposes a methodology based on the Fuzzy TOPSIS method and the PCA to hierarchize and group the identified critical factors.

3.3 METHODS AND RESULTS

3.3.1 Hierarchy for Evaluation

The analysis of the critical factors for the implementation of ITG made it possible to identify which ones are considered relevant and thus to make decisions to align strategic objectives and IT. In this research, a fuzzy multicriteria decision-making method was implemented for the analysis and classification of critical factors of ITG in an organizational environment based on the evaluation of 4 ITG experts. The ITG areas proposed by the Organization for Economic Cooperation and Development (OECD, 2021), as well as numerous critical factors for ITG implementation (alternatives) were considered as criteria in this topic. The results of this study are used to prioritize the critical factors in order of importance from the most critical to the least crucial. The scoring of the critical factors based on their importance informs decision makers which key areas are of highest priority for resource investment. Based on the literature review and expert assessments, we have identified 68 critical factors for ITG implementation, categorized within Grembergen's (Van Grembergen et al., 2004) proposed ITG mechanisms, structure, processes, and relational mechanisms. The Fuzzy TOPSIS method alternatives were named [A1-A19], [A20-A51], and [A52-A68] for the alternatives according to the classification and [E1-E19], [P20-P51], and [M52-M68] in PCA for the items corresponding to each classification. To prioritize the critical factors we have recognized as criteria, the seven key areas of ITG indicated by the OECD as strategic alignment, IT value and performance, IT risk management, transparency and accountability, compliance, IT resource management, innovation and continuous improvement denoted by C1, C2, C3, C4, C5, C6, and C7, respectively.

3.3.2 Fuzzy Topsis Method

The Fuzzy TOPSIS (Fuzzy Technique for Order Preference by Similarity to Ideal Solution) multicriteria decision-making method was selected as a tool to hierarchize the critical variables associated with the successful implementation of the ITG. This choice responds to the need to address a complex and multidimensional decision problem, characterized by the presence of multiple evaluation criteria (the key areas of governance) and high levels of uncertainty and subjectivity in the expert judgments, which were expressed through linguistic terms (Azhar et al., 2021; Ramirez-Ochoa et al., 2022; Kumar & Pamucar, 2025).

The Fuzzy TOPSIS procedure was developed by Chen and Hwang (1992), who extended the classical TOPSIS method by incorporating fuzzy numbers to capture the ambiguity inherent in human judgments (Schmitz & Pape, 2020). Subsequently, Chen came up with the use of the vertex method to compute the distance between two triangular fuzzy numbers (TFNs), consolidating its application in environments with vague information (Bruma, 2020). TFNs have proven to be effective in representing imprecise judgments and adequately modeling the uncertainty inherent in most decision-making problems (Fathi et al., 2011).

Fuzzy TOPSIS allows the establishment of a positive ideal solution and a negative anti-ideal solution, which serve as benchmarks to determine the relative closeness of each alternative. This methodology integrates fuzzy logic to convert linguistic valuations into mathematical representations, thus facilitating the comparison, prioritization, and ranking of alternatives under conditions of ambiguity (Salih et al., 2019; Ayağ, 2021; Wang, 2022; Sahoo & Goswami, 2023; Kumar & Pamucar, 2025).

In this research 68 alternatives and 7 criteria corresponding to critical factors for ITG implementation and key areas of ITG proposed by OECD (OECD, 2021) respectively, were consistently assessed using the Fuzzy TOPSIS procedure. The criteria weights and evaluations of each of the alternatives were generated using the ratings of the ITG decision-making experts. The following are the six methodological steps that were followed to rank the alternatives.

Step 1: Create a combined decision matrix.

Because natural language is imprecise, the fuzzy model allowed representing that imprecision using TFN. The letters *l, m,* and *u* are used to demonstrate a TFN. The linguistic terms of the evaluation scale were replaced by the associated fuzzy numbers as illustrated in the standard fuzzy scale in Table 3.1.

TABLE 3.1
Fuzzy Scale

Code	Linguistic Terms	Fuzzy Number (*l, m, u*)
1	None at all	1, 1, 2
2	Very little	1, 2, 3
3	Very little	2, 3, 4
4	Moderately	3, 4, 5
5	Fairly	4, 5, 6
6	Very much	5, 6, 7
7	To a great extent	6, 7, 7

This subsequently allowed the relative importance of each criterion to be calculated mathematically and facilitated the weighting of the decision matrix by applying the fuzzy weights to the evaluations and making the model closer to reality, capturing the uncertainty and ambiguity inherent in human judgments and human language (Ayağ, 2021).

A decision matrix consists of a set of values in columns and rows used to compare different alternatives by weighting parameters according to their importance. The decision matrix combined with the arithmetic mean of the evaluators' opinions is shown in Table 3.2. The evaluations of all alternatives against each criterion are expressed as fuzzy numbers using the equations illustrated in (3.1).

$$\tilde{x}_{ij} = \left(a_{ij}, b_{ij}, c_{ij}\right) a_{ij} = \min_{k}\left\{a_{ij}^{k}\right\} b_{ij} = \frac{1}{K}\sum_{k=1}^{K} b_{ij}^{k} c_{ij} = \max_{k}\left\{c_{ij}^{k}\right\} \qquad (3.1)$$

The combined decision matrix unifies the evaluation criteria of several experts into a single matrix that will be the basis of the following steps of the Fuzzy TOPSIS method such as normalization, weighting, and calculation of distances to the ideal and anti-ideal solutions. It finally ensures that decisions are not based on the opinion of a single expert, but on a fuzzy consensus.

TABLE 3.2
Combined Decision Matrix

	C1	C2	C3	C4	C5	C6	C7
A1	6 , 7.0 , 7	5 , 6.75 , 7	2 , 5 , 7	3 , 5.75 , 7	3 , 4.25 , 6	4 , 5.75 , 7	3 , 5.75 , 7
A2	2 , 4.0 , 6	2 , 4.5 , 7	2 , 4.25 , 6	2 , 4.75 , 7	1 , 3.75 , 7	1 , 3.5 , 6	3 , 5 , 7
A3	4 , 6.5 , 7	4 , 5.75 , 7	4 , 5 , 6	2 , 5.75 , 7	2 , 5.5 , 7	4 , 6.25 , 7	4 , 5.75 , 7
A4	4 , 6.3 , 7	4 , 6 , 7	3 , 4.75 , 7	5 , 6.5 , 7	4 , 6 , 7	3 , 6 , 7	3 , 5.75 , 7
A5	4 , 6.0 , 7	3 , 5.25 , 7	3 , 4.75 , 7	4 , 6.25 , 7	4 , 5.75 , 7	3 , 5.75 , 7	3 , 5.75 , 7
A6	4 , 5.8 , 7	3 , 4.25 , 6	3 , 5.75 , 7	3 , 6 , 7	3 , 5.25 , 7	3 , 4.75 , 6	2 , 4 , 6
A7	4 , 6.0 , 7	3 , 5.75 , 7	2 , 4.25 , 7	2 , 4.5 , 7	2 , 3.75 , 5	3 , 5 , 7	4 , 5.75 , 7
A8	2 , 6.0 , 7	2 , 5.75 , 7	2 , 5.25 , 7	2 , 5.75 , 7	2 , 5 , 7	2 , 5.75 , 7	2 , 5.75 , 7
A9	4 , 6.3 , 7	5 , 6.25 , 7	3 , 4.75 , 7	3 , 5.25 , 7	3 , 5.25 , 7	3 , 6 , 7	2 , 5.25 , 7
A10	3 , 6.0 , 7	3 , 6 , 7	3 , 5 , 7	2 , 5 , 7	2 , 5 , 7	2 , 5.75 , 7	1 , 5.5 , 7
A11	4 , 6.5 , 7	3 , 5.5 , 7	3 , 5.75 , 7	3 , 6 , 7	3 , 5.75 , 7	3 , 5.75 , 7	3 , 5 , 7
A12	3 , 5.8 , 7	3 , 5.75 , 7	3 , 5.75 , 7	3 , 5.75 , 7	3 , 5.5 , 7	3 , 5.75 , 7	3 , 5.25 , 7
A13	4 , 6.0 , 7	4 , 6.5 , 7	2 , 4.5 , 6	3 , 6 , 7	3 , 6 , 7	3 , 6 , 7	2 , 5.75 , 7
A14	4 , 6.3 , 7	3 , 5.5 , 7	3 , 5 , 7	2 , 5 , 7	1 , 5 , 7	2 , 5.75 , 7	3 , 5.75 , 7
A15	3 , 5.8 , 7	3 , 6 , 7	2 , 4.75 , 7	2 , 5.75 , 7	3 , 5.75 , 7	3 , 6 , 7	3 , 6 , 7
A16	3 , 5.0 , 7	3 , 5 , 7	4 , 5.75 , 7	4 , 6 , 7	4 , 6.5 , 7	3 , 5.25 , 7	2 , 4 , 7
A17	4 , 5.3 , 7	4 , 6.25 , 7	3 , 5.5 , 7	2 , 4.75 , 7	2 , 4.25 , 7	3 , 5.5 , 7	3 , 6 , 7
A18	3 , 5.5 , 7	4 , 6.25 , 7	4 , 5.75 , 7	3 , 5.5 , 7	2 , 4.75 , 7	3 , 5.75 , 7	4 , 6 , 7
A19	4 , 6.3 , 7	4 , 6.25 , 7	3 , 5.5 , 7	3 , 5.5 , 7	3 , 5.25 , 7	4 , 6 , 7	4 , 6.25 , 7
A20	2 , 5.3 , 7	3 , 5.5 , 7	3 , 5.5 , 7	3 , 5 , 7	2 , 4.25 , 7	4 , 5.75 , 7	4 , 6 , 7
A21	5 , 6.5 , 7	4 , 6 , 7	3 , 5.25 , 7	3 , 5.5 , 7	2 , 5 , 7	3 , 5.25 , 7	3 , 5.25 , 7

(Continued)

TABLE 3.2 *(Continued)*
Combined Decision Matrix

	C1	C2	C3	C4	C5	C6	C7
A22	2 , 5.0 , 7	3 , 5.25 , 7	4 , 6.25 , 7	4 , 5.5 , 7	3 , 5.25 , 7	3 , 5.5 , 7	3 , 4.25 , 6
A23	3 , 5.8 , 7	4 , 5.5 , 7	3 , 5.5 , 7	2 , 5 , 7	3 , 5.25 , 7	4 , 6 , 7	3 , 5 , 7
A24	1 , 3.3 , 6	1 , 3.5 , 7	1 , 3.5 , 6	1 , 4.25 , 7	3 , 5 , 7	3 , 4 , 5	2 , 4.25 , 6
A25	4 , 6.3 , 7	4 , 5.75 , 7	3 , 5.25 , 7	4 , 6 , 7	3 , 5.5 , 7	3 , 5 , 7	2 , 3.75 , 6
A26	2 , 4.5 , 7	3 , 4.5 , 6	4 , 5.75 , 7	3 , 5.75 , 7	4 , 6 , 7	3 , 6 , 7	2 , 5.5 , 7
A27	4 , 6.5 , 7	3 , 5.75 , 7	3 , 4.25 , 6	2 , 4 , 6	2 , 4 , 6	2 , 5.25 , 7	3 , 4.75 , 7
A28	4 , 6.0 , 7	3 , 5.75 , 7	3 , 5.25 , 7	2 , 5.25 , 7	2 , 5 , 7	2 , 5.5 , 7	2 , 5.5 , 7
A29	4 , 6.0 , 7	3 , 5.5 , 7	3 , 5.5 , 7	4 , 5.5 , 7	3 , 5.5 , 7	3 , 5.75 , 7	3 , 5.75 , 7
A30	4 , 6.3 , 7	2 , 5.5 , 7	3 , 5.25 , 7	3 , 5.75 , 7	2 , 5.5 , 7	3 , 5.75 , 7	2 , 5.5 , 7
A31	3 , 5.5 , 7	2 , 4.75 , 7	3 , 5.25 , 7	2 , 4.75 , 7	2 , 4 , 6	2 , 4.25 , 6	3 , 5.25 , 7
A32	3 , 5.5 , 7	4 , 6.25 , 7	2 , 4.25 , 6	2 , 4.5 , 7	2 , 4 , 6	2 , 5.5 , 7	2 , 5.25 , 7
A33	2 , 4.0 , 6	3 , 5.25 , 7	3 , 5.25 , 7	3 , 4.75 , 7	3 , 5 , 7	3 , 5.5 , 7	3 , 5.75 , 7
A34	1 , 4.3 , 7	3 , 4.75 , 7	4 , 6.25 , 7	3 , 5 , 7	3 , 5.25 , 7	3 , 5 , 7	3 , 4.75 , 7
A35	2 , 4.0 , 7	2 , 4.5 , 7	4 , 6 , 7	3 , 5 , 7	2 , 4.75 , 7	4 , 5.25 , 7	2 , 3.75 , 6
A36	2 , 4.0 , 7	3 , 4.75 , 7	4 , 6 , 7	3 , 5 , 7	3 , 5 , 7	3 , 5 , 7	2 , 4.25 , 6
A37	2 , 4.0 , 7	3 , 4.75 , 7	4 , 6.25 , 7	2 , 4.5 , 7	3 , 5 , 7	2 , 5 , 7	2 , 3.75 , 6
A38	2 , 4.8 , 7	3 , 5.25 , 7	4 , 6 , 7	3 , 4.25 , 6	3 , 4.75 , 7	3 , 5.25 , 7	3 , 4.75 , 7
A39	2 , 4.3 , 7	2 , 4.25 , 7	2 , 5.25 , 7	3 , 5.5 , 7	3 , 5.75 , 7	3 , 5.25 , 7	2 , 3.75 , 6
A40	3 , 5.3 , 7	2 , 5.25 , 7	3 , 4.75 , 7	2 , 4.25 , 6	2 , 4 , 6	2 , 5.5 , 7	3 , 5.75 , 7
A41	3 , 5.0 , 7	2 , 5.25 , 7	3 , 4.75 , 7	2 , 4.25 , 6	2 , 4 , 6	2 , 5.5 , 7	3 , 5.75 , 7
A42	3 , 5.8 , 7	3 , 5.75 , 7	4 , 6 , 7	2 , 5.5 , 7	2 , 5.5 , 7	3 , 5.75 , 7	3 , 5.75 , 7
A43	3 , 5.8 , 7	4 , 5.75 , 7	4 , 6 , 7	2 , 4.5 , 7	2 , 4.5 , 7	3 , 5.5 , 7	3 , 5.75 , 7

A44	3 , 6.0 , 7	3 , 5.5 , 7	2 , 5.5 , 7	2 , 4.5 , 7	2 , 4.25 , 6	3 , 5.75 , 7	4 , 6 , 7
A45	2 , 5.8 , 7	3 , 5 , 7	5 , 6.5 , 7	2 , 5 , 7	3 , 5.5 , 7	3 , 5.5 , 7	2 , 5.25 , 7
A46	2 , 5.5 , 7	3 , 5.25 , 7	3 , 5.5 , 7	2 , 5 , 7	3 , 5.5 , 7	3 , 5.25 , 7	3 , 4.5 , 6
A47	2 , 4.3 , 7	3 , 4.75 , 7	4 , 6.25 , 7	2 , 4.75 , 7	2 , 5.25 , 7	2 , 5 , 7	2 , 4.25 , 6
A48	3 , 4.5 , 7	4 , 5 , 6	4 , 5.75 , 7	2 , 4.75 , 7	2 , 4.75 , 7	3 , 5.25 , 7	3 , 4.75 , 7
A49	2 , 4.5 , 7	3 , 5.75 , 7	4 , 6 , 7	2 , 5.5 , 7	3 , 5.75 , 7	2 , 5.5 , 7	2 , 4.5 , 7
A50	2 , 5.0 , 7	4 , 6 , 7	2 , 4 , 6	2 , 5.5 , 7	2 , 4 , 6	3 , 5.75 , 7	3 , 5.75 , 7
A51	4 , 6.0 , 7	4 , 6 , 7	3 , 4.25 , 6	3 , 5.75 , 7	3 , 4.25 , 6	4 , 6 , 7	2 , 5.5 , 7
A52	4 , 6.3 , 7	3 , 5.5 , 7	2 , 4.25 , 6	4 , 6 , 7	3 , 4.75 , 7	3 , 5.75 , 7	3 , 5.75 , 7
A53	4 , 6.5 , 7	4 , 6 , 7	3 , 5.5 , 7	3 , 5.75 , 7	3 , 5.5 , 7	4 , 6 , 7	4 , 6 , 7
A54	4 , 6.3 , 7	4 , 6 , 7	3 , 5 , 7	4 , 6 , 7	2 , 5.5 , 7	3 , 5.75 , 7	3 , 5.75 , 7
A55	5 , 6.5 , 7	3 , 5.75 , 7	4 , 5.25 , 7	3 , 5.75 , 7	3 , 5.25 , 7	3 , 5.75 , 7	2 , 4.75 , 7
A56	3 , 5.8 , 7	3 , 5.75 , 7	3 , 5 , 7	3 , 5.5 , 7	3 , 5.5 , 7	3 , 5.75 , 7	3 , 5.75 , 7
A57	5 , 6.5 , 7	3 , 5.75 , 7	2 , 4.5 , 6	4 , 5.25 , 7	4 , 5.5 , 7	3 , 5.5 , 7	4 , 6 , 7
A58	5 , 6.3 , 7	3 , 5.5 , 7	3 , 4.25 , 6	4 , 5.75 , 7	3 , 5.25 , 7	3 , 5.5 , 7	3 , 5.5 , 7
A59	4 , 6.0 , 7	4 , 6 , 7	4 , 5.5 , 7	3 , 5 , 7	3 , 4.5 , 6	3 , 5.75 , 7	3 , 5.5 , 7
A60	3 , 5.5 , 7	4 , 6 , 7	3 , 5.25 , 7	3 , 5.75 , 7	4 , 5.5 , 7	3 , 5.75 , 7	3 , 5 , 7
A61	2 , 4.8 , 7	2 , 4.25 , 6	3 , 5 , 7	3 , 5.5 , 7	4 , 5.75 , 7	3 , 5.5 , 7	2 , 4.5 , 7
A62	2 , 5.5 , 7	2 , 4.5 , 7	3 , 5.25 , 7	3 , 5.75 , 7	4 , 6 , 7	3 , 5.75 , 7	2 , 4.75 , 7
A63	1 , 3.8 , 6	1 , 3.75 , 6	3 , 5 , 7	4 , 6 , 7	4 , 6 , 7	2 , 4.5 , 7	2 , 3.75 , 6
A64	3 , 5.3 , 7	2 , 4.75 , 7	3 , 5.25 , 7	3 , 5.25 , 7	4 , 5.5 , 7	3 , 5 , 7	2 , 4 , 6
A65	2 , 4.0 , 7	3 , 5 , 7	3 , 5.75 , 7	4 , 5.5 , 7	4 , 5.75 , 7	3 , 5.5 , 7	2 , 4 , 6
A66	3 , 6.0 , 7	3 , 5.75 , 7	3 , 5.25 , 7	3 , 5.75 , 7	3 , 6 , 7	2 , 5.5 , 7	3 , 5.5 , 7
A67	4 , 6.0 , 7	4 , 5.75 , 7	3 , 5.5 , 7	3 , 5.5 , 7	3 , 5.75 , 7	3 , 5 , 7	2 , 4.5 , 7
A68	3 , 5.5 , 7	3 , 5.5 , 7	4 , 5.75 , 7	3 , 5.5 , 7	3 , 5.25 , 7	3 , 5.25 , 7	2 , 4.25 , 6

Step 2: Develop a fuzzy normalized decision matrix.

From the fuzzy decision matrix, the highest value for each criterion is identified.

This value is used to normalize the fuzzy numbers associated with the criterion C* and the normalization formula (3.2) is applied for beneficial criteria where higher is better.

$$\tilde{r}_{ij} = \left(\frac{a_{ij}}{c_j^*}, \frac{b_{ij}}{c_j^*}, \frac{c_{ij}}{c_j^*} \right) \qquad c_j^* = {}^{\max}_{i}(c_{ij}) \tag{3.2}$$

To normalize the fuzzy numbers associated with criterion C, the normalization formula (3.3) is applied for non-beneficial criteria where lower is better.

$$\tilde{r}_{ij} = \left(\frac{a_j^-}{c_{ij}}, \frac{a_j^-}{b_{ij}}, \frac{a_j^-}{a_{ij}} \right) \qquad a_j^- = {}^{\min}_{i}(a_{ij}) \tag{3.3}$$

Once the values and $c_j^* a_j^-$ are obtained, the normalized values are still TFNs, and the matrix generated in this step is illustrated in Table 3.3 will serve as the basis for Step 3, calculation of the fuzzy weighted normalized decision matrix.

Step 3: Make a fuzzy-weighted normalized decision matrix.

As the linguistic term ratings were defined in fuzzy terms l_{ij}, m_{ij}, u_{ij} the calculation of the weights of each of the seven criteria was obtained using TFNs and the fuzzy weight average method. As a first step in the process, the triangular fuzzy average was calculated for each criterion with the formula (3.4).

$$\tilde{w}_j = \left(\frac{1}{n} \sum_{i=1}^{n} l_{ij}, \frac{1}{n} \sum_{i=1}^{n} m_{ij}, \frac{1}{n} \sum_{i=1}^{n} u_{ij} \right) \tag{3.4}$$

where:

$\tilde{w}_j$ (l(j,) m(j), uj) is the fuzzy weight of criterion j.

$l_{(ij,)\ m(ij)}$, u_{ij} are the ratings of evaluator i for criterion j.

n is the number of evaluators.

The second part consisted in normalizing the TFN fuzzy weights of each criterion so that the weights can be compared and add up to approximately 1. The normalization is done by dividing each component l, m, l, by the total sum of each respective component for all criteria (3.5).

$$\tilde{w}_j^{norm} = \left(\frac{l_j}{\sum_{k=1}^{m} l_k}, \frac{m_j}{\sum_{k=1}^{m} m_k}, \frac{u_j}{\sum_{k=1}^{m} u_k} \right) \tag{3.5}$$

where:

m is the total number of criteria.

TABLE 3.3
Fuzzy Normalized Decision Matrix

	C1	C2	C3	C4	C5	C6	C7
A1	0.8571, 1.0000, 1.0000	0.7143, 0.9643, 1.0000	0.2857, 0.7143, 1.0000	0.0000, 0.8214, 1.0000	0.4286, 0.6071, 0.8571	0.5714, 0.8214, 1.0000	0.4286, 0.8214, 1.0000
A2	0.2857, 0.5714, 0.8571	0.2857, 0.6429, 1.0000	0.2857, 0.6071, 0.8571	0.2857, 0.6786, 1.0000	0.1429, 0.5357, 1.0000	0.1429, 0.5000, 0.8571	0.4286, 0.7143, 1.0000
A3	0.5714, 0.9286, 1.0000	0.5714, 0.8214, 1.0000	0.5714, 0.7143, 0.8571	0.2857, 0.8214, 1.0000	0.2857, 0.7857, 1.0000	0.5714, 0.8929, 1.0000	0.5714, 0.8214, 1.0000
A4	0.5714, 0.8929, 1.0000	0.5714, 0.8571, 1.0000	0.4286, 0.6786, 1.0000	0.7143, 0.9286, 1.0000	0.5714, 0.8571, 1.0000	0.4286, 0.8571, 1.0000	0.4286, 0.8214, 1.0000
A5	0.5714, 0.8571, 1.0000	0.4286, 0.7500, 1.0000	0.4286, 0.6786, 1.0000	0.5714, 0.8929, 1.0000	0.5714, 0.8214, 1.0000	0.4286, 0.8214, 1.0000	0.4286, 0.8214, 1.0000
A6	0.5714, 0.8214, 1.0000	0.4286, 0.6071, 0.8571	0.4286, 0.8214, 1.0000	0.4286, 0.8571, 1.0000	0.4286, 0.7500, 1.0000	0.4286, 0.6786, 0.8571	0.2857, 0.5714, 0.8571
A7	0.5714, 0.8571, 1.0000	0.4286, 0.8214, 1.0000	0.2857, 0.6071, 1.0000	0.2857, 0.6429, 1.0000	0.2857, 0.5357, 0.7143	0.4286, 0.7143, 1.0000	0.5714, 0.8214, 1.0000
A8	0.2857, 0.8571, 1.0000	0.2857, 0.8214, 1.0000	0.2857, 0.7500, 1.0000	0.2857, 0.8214, 1.0000	0.2857, 0.7143, 1.0000	0.2857, 0.8214, 1.0000	0.2857, 0.8214, 1.0000
A9	0.5714, 0.8929, 1.0000	0.7143, 0.8929, 1.0000	0.4286, 0.6786, 1.0000	0.4286, 0.7500, 1.0000	0.4286, 0.7500, 1.0000	0.4286, 0.8571, 1.0000	0.2857, 0.7500, 1.0000
A10	0.4286, 0.8571, 1.0000	0.4286, 0.8571, 1.0000	0.4286, 0.7143, 1.0000	0.2857, 0.7143, 1.0000	0.2857, 0.7143, 1.0000	0.2857, 0.8214, 1.0000	0.1429, 0.7857, 1.0000
A11	0.5714, 0.9286, 1.0000	0.4286, 0.7857, 1.0000	0.4286, 0.8214, 1.0000	0.4286, 0.8571, 1.0000	0.4286, 0.8214, 1.0000	0.4286, 0.8214, 1.0000	0.4286, 0.7143, 1.0000
A12	0.4286, 0.8214, 1.0000	0.4286, 0.8214, 1.0000	0.4286, 0.8214, 1.0000	0.4286, 0.8214, 1.0000	0.4286, 0.7857, 1.0000	0.4286, 0.8214, 1.0000	0.4286, 0.7500, 1.0000
A13	0.5714, 0.8571, 1.0000	0.5714, 0.9286, 1.0000	0.2857, 0.6429, 0.8571	0.4286, 0.8571, 1.0000	0.4286, 0.8571, 1.0000	0.4286, 0.8571, 1.0000	0.2857, 0.8214, 1.0000
A14	0.5714, 0.8929, 1.0000	0.4286, 0.7857, 1.0000	0.4286, 0.7143, 1.0000	0.2857, 0.7143, 1.0000	0.1429, 0.7143, 1.0000	0.2857, 0.8214, 1.0000	0.4286, 0.8214, 1.0000
A15	0.4286, 0.8214, 1.0000	0.4286, 0.8571, 1.0000	0.2857, 0.6786, 1.0000	0.2857, 0.8214, 1.0000	0.4286, 0.8214, 1.0000	0.4286, 0.8571, 1.0000	0.4286, 0.8571, 1.0000
A16	0.4286, 0.7143, 1.0000	0.4286, 0.7143, 1.0000	0.5714, 0.8214, 1.0000	0.5714, 0.8571, 1.0000	0.5714, 0.9286, 1.0000	0.4286, 0.7500, 1.0000	0.2857, 0.5714, 1.0000
A17	0.5714, 0.7500, 1.0000	0.5714, 0.8929, 1.0000	0.4286, 0.7857, 1.0000	0.2857, 0.6786, 1.0000	0.2857, 0.6071, 1.0000	0.4286, 0.7857, 1.0000	0.4286, 0.8571, 1.0000
A18	0.4286, 0.7857, 1.0000	0.5714, 0.8929, 1.0000	0.5714, 0.8214, 1.0000	0.4286, 0.7857, 1.0000	0.2857, 0.6786, 1.0000	0.4286, 0.8214, 1.0000	0.5714, 0.8571, 1.0000
A19	0.5714, 0.8929, 1.0000	0.5714, 0.8929, 1.0000	0.4286, 0.7857, 1.0000	0.4286, 0.7857, 1.0000	0.4286, 0.7500, 1.0000	0.5714, 0.8571, 1.0000	0.5714, 0.8929, 1.0000
A20	0.2857, 0.7500, 1.0000	0.4286, 0.7857, 1.0000	0.4286, 0.7857, 1.0000	0.4286, 0.7143, 1.0000	0.2857, 0.6071, 1.0000	0.5714, 0.8214, 1.0000	0.5714, 0.8571, 1.0000
A21	0.7143, 0.9286, 1.0000	0.5714, 0.8571, 1.0000	0.4286, 0.7500, 1.0000	0.4286, 0.7857, 1.0000	0.2857, 0.7143, 1.0000	0.4286, 0.7500, 1.0000	0.4286, 0.7500, 1.0000

(Continued)

TABLE 3.3 *(Continued)*
Fuzzy Normalized Decision Matrix

	C1	C2	C3	C4	C5	C6	C7
A22	0.2857, 0.7143, 1.0000	0.4286, 0.7500, 1.0000	0.5714, 0.8929, 1.0000	0.5714, 0.7857, 1.0000	0.4286, 0.7500, 1.0000	0.4286, 0.7857, 1.0000	0.4286, 0.6071, 0.8571
A23	0.4286, 0.8214, 1.0000	0.5714, 0.7857, 1.0000	0.4286, 0.7857, 1.0000	0.2857, 0.7143, 1.0000	0.4286, 0.7500, 1.0000	0.5714, 0.8571, 1.0000	0.4286, 0.7143, 1.0000
A24	0.1429, 0.4643, 0.8571	0.1429, 0.5000, 1.0000	0.1429, 0.5000, 0.8571	0.1429, 0.6071, 1.0000	0.4286, 0.7143, 1.0000	0.4286, 0.5714, 0.7143	0.2857, 0.6071, 0.8571
A25	0.5714, 0.8929, 1.0000	0.5714, 0.8214, 1.0000	0.4286, 0.7500, 1.0000	0.5714, 0.8571, 1.0000	0.4286, 0.7857, 1.0000	0.4286, 0.7143, 1.0000	0.2857, 0.5357, 0.8571
A26	0.2857, 0.6429, 1.0000	0.4286, 0.6429, 0.8571	0.5714, 0.8214, 1.0000	0.4286, 0.8214, 1.0000	0.5714, 0.8571, 1.0000	0.4286, 0.8571, 1.0000	0.2857, 0.7857, 1.0000
A27	0.5714, 0.9286, 1.0000	0.4286, 0.8214, 1.0000	0.4286, 0.6071, 0.8571	0.2857, 0.5714, 0.8571	0.2857, 0.5714, 0.8571	0.2857, 0.7500, 1.0000	0.4286, 0.6786, 1.0000
A28	0.5714, 0.8571, 1.0000	0.4286, 0.8214, 1.0000	0.4286, 0.7500, 1.0000	0.2857, 0.7500, 1.0000	0.2857, 0.7143, 1.0000	0.2857, 0.7857, 1.0000	0.2857, 0.7857, 1.0000
A29	0.5714, 0.8571, 1.0000	0.4286, 0.7857, 1.0000	0.4286, 0.7857, 1.0000	0.5714, 0.7857, 1.0000	0.4286, 0.7857, 1.0000	0.4286, 0.8214, 1.0000	0.4286, 0.8214, 1.0000
A30	0.5714, 0.8929, 1.0000	0.2857, 0.7857, 1.0000	0.4286, 0.7500, 1.0000	0.4286, 0.8214, 1.0000	0.2857, 0.7857, 1.0000	0.4286, 0.8214, 1.0000	0.2857, 0.7857, 1.0000
A31	0.4286, 0.7857, 1.0000	0.2857, 0.6786, 1.0000	0.4286, 0.7500, 1.0000	0.2857, 0.6786, 1.0000	0.2857, 0.5714, 0.8571	0.2857, 0.6071, 0.8571	0.4286, 0.7500, 1.0000
A32	0.4286, 0.7857, 1.0000	0.5714, 0.8929, 1.0000	0.2857, 0.6071, 0.8571	0.2857, 0.6429, 1.0000	0.2857, 0.5714, 0.8571	0.2857, 0.7857, 1.0000	0.2857, 0.7500, 1.0000
A33	0.2857, 0.5714, 0.8571	0.4286, 0.7500, 1.0000	0.4286, 0.7500, 1.0000	0.4286, 0.6786, 1.0000	0.4286, 0.7143, 1.0000	0.4286, 0.7857, 1.0000	0.4286, 0.8214, 1.0000
A34	0.1429, 0.6071, 1.0000	0.4286, 0.6786, 1.0000	0.5714, 0.8929, 1.0000	0.4286, 0.7143, 1.0000	0.4286, 0.7500, 1.0000	0.4286, 0.7143, 1.0000	0.4286, 0.6786, 1.0000
A35	0.2857, 0.5714, 1.0000	0.2857, 0.6429, 1.0000	0.5714, 0.8571, 1.0000	0.4286, 0.7143, 1.0000	0.2857, 0.6786, 1.0000	0.5714, 0.7500, 1.0000	0.2857, 0.5357, 0.8571
A36	0.2857, 0.5714, 1.0000	0.4286, 0.6786, 1.0000	0.5714, 0.8571, 1.0000	0.4286, 0.7143, 1.0000	0.4286, 0.7143, 1.0000	0.4286, 0.7143, 1.0000	0.2857, 0.6071, 0.8571
A37	0.2857, 0.5714, 1.0000	0.4286, 0.6786, 1.0000	0.5714, 0.8929, 1.0000	0.2857, 0.6429, 1.0000	0.4286, 0.7143, 1.0000	0.2857, 0.7143, 1.0000	0.2857, 0.5357, 0.8571
A38	0.2857, 0.6786, 1.0000	0.4286, 0.7500, 1.0000	0.5714, 0.8571, 1.0000	0.4286, 0.6071, 0.8571	0.4286, 0.6786, 1.0000	0.4286, 0.7500, 1.0000	0.4286, 0.6786, 1.0000
A39	0.2857, 0.6071, 1.0000	0.2857, 0.6071, 1.0000	0.2857, 0.7500, 1.0000	0.4286, 0.7857, 1.0000	0.4286, 0.8214, 1.0000	0.4286, 0.7500, 1.0000	0.2857, 0.5357, 0.8571
A40	0.4286, 0.7500, 1.0000	0.2857, 0.7500, 1.0000	0.4286, 0.6786, 1.0000	0.2857, 0.6071, 0.8571	0.2857, 0.5714, 0.8571	0.2857, 0.7857, 1.0000	0.4286, 0.8214, 1.0000
A41	0.4286, 0.7143, 1.0000	0.2857, 0.7500, 1.0000	0.4286, 0.6786, 1.0000	0.2857, 0.6071, 0.8571	0.2857, 0.5714, 0.8571	0.2857, 0.7857, 1.0000	0.4286, 0.8214, 1.0000
A42	0.4286, 0.8214, 1.0000	0.4286, 0.8214, 1.0000	0.5714, 0.8571, 1.0000	0.2857, 0.7857, 1.0000	0.2857, 0.7857, 1.0000	0.4286, 0.8214, 1.0000	0.4286, 0.8214, 1.0000
A43	0.4286, 0.8214, 1.0000	0.5714, 0.8214, 1.0000	0.5714, 0.8571, 1.0000	0.2857, 0.6429, 1.0000	0.2857, 0.6429, 1.0000	0.4286, 0.7857, 1.0000	0.4286, 0.8214, 1.0000
A44	0.4286, 0.8571, 1.0000	0.4286, 0.7857, 1.0000	0.2857, 0.7857, 1.0000	0.2857, 0.6429, 1.0000	0.2857, 0.6071, 0.8571	0.4286, 0.8214, 1.0000	0.5714, 0.8571, 1.0000

A45	0.2857, 0.8214, 1.0000	0.4286, 0.7143, 1.0000	0.7143, 0.9286, 1.0000	0.2857, 0.7143, 1.0000	0.4286, 0.7857, 1.0000	0.4286, 0.7857, 1.0000	0.2857, 0.7500, 1.0000
A46	0.2857, 0.7857, 1.0000	0.4286, 0.7500, 1.0000	0.4286, 0.7857, 1.0000	0.2857, 0.7143, 1.0000	0.4286, 0.7857, 1.0000	0.4286, 0.7500, 1.0000	0.4286, 0.6429, 0.8571
A47	0.2857, 0.6071, 1.0000	0.4286, 0.6786, 1.0000	0.5714, 0.8929, 1.0000	0.2857, 0.6786, 1.0000	0.2857, 0.7500, 1.0000	0.2857, 0.7143, 1.0000	0.2857, 0.6071, 0.8571
A48	0.4286, 0.6429, 1.0000	0.5714, 0.7143, 0.8571	0.5714, 0.8214, 1.0000	0.2857, 0.6786, 1.0000	0.2857, 0.6786, 1.0000	0.4286, 0.7500, 1.0000	0.4286, 0.6786, 1.0000
A49	0.2857, 0.6429, 1.0000	0.4286, 0.8214, 1.0000	0.5714, 0.8571, 1.0000	0.2857, 0.7857, 1.0000	0.4286, 0.8214, 1.0000	0.2857, 0.7857, 1.0000	0.2857, 0.6429, 1.0000
A50	0.2857, 0.7143, 1.0000	0.5714, 0.8571, 1.0000	0.2857, 0.5714, 0.8571	0.2857, 0.7857, 1.0000	0.2857, 0.5714, 0.8571	0.4286, 0.8214, 1.0000	0.4286, 0.8214, 1.0000
A51	0.5714, 0.8571, 1.0000	0.5714, 0.8571, 1.0000	0.4286, 0.6071, 0.8571	0.4286, 0.8214, 1.0000	0.4286, 0.6071, 0.8571	0.5714, 0.8571, 1.0000	0.2857, 0.7857, 1.0000
A52	0.5714, 0.8929, 1.0000	0.4286, 0.7857, 1.0000	0.2857, 0.6071, 0.8571	0.5714, 0.8571, 1.0000	0.4286, 0.6786, 1.0000	0.4286, 0.8214, 1.0000	0.4286, 0.8214, 1.0000
A53	0.5714, 0.9286, 1.0000	0.5714, 0.8571, 1.0000	0.4286, 0.7857, 1.0000	0.4286, 0.8214, 1.0000	0.4286, 0.7857, 1.0000	0.5714, 0.8571, 1.0000	0.5714, 0.8571, 1.0000
A54	0.5714, 0.8929, 1.0000	0.5714, 0.8571, 1.0000	0.4286, 0.7143, 1.0000	0.5714, 0.8571, 1.0000	0.2857, 0.7857, 1.0000	0.4286, 0.8214, 1.0000	0.4286, 0.8214, 1.0000
A55	0.7143, 0.9286, 1.0000	0.4286, 0.8214, 1.0000	0.5714, 0.7500, 1.0000	0.4286, 0.8214, 1.0000	0.4286, 0.7500, 1.0000	0.4286, 0.8214, 1.0000	0.2857, 0.6786, 1.0000
A56	0.4286, 0.8214, 1.0000	0.4286, 0.8214, 1.0000	0.4286, 0.7143, 1.0000	0.4286, 0.7857, 1.0000	0.4286, 0.7857, 1.0000	0.4286, 0.8214, 1.0000	0.4286, 0.8214, 1.0000
A57	0.7143, 0.9286, 1.0000	0.4286, 0.8214, 1.0000	0.2857, 0.6429, 0.8571	0.5714, 0.7500, 1.0000	0.5714, 0.7857, 1.0000	0.4286, 0.7857, 1.0000	0.5714, 0.8571, 1.0000
A58	0.7143, 0.8929, 1.0000	0.4286, 0.7857, 1.0000	0.4286, 0.6071, 0.8571	0.5714, 0.8214, 1.0000	0.4286, 0.7500, 1.0000	0.4286, 0.7857, 1.0000	0.4286, 0.7857, 1.0000
A59	0.5714, 0.8571, 1.0000	0.5714, 0.8571, 1.0000	0.5714, 0.7857, 1.0000	0.4286, 0.7143, 1.0000	0.4286, 0.6429, 0.8571	0.4286, 0.8214, 1.0000	0.4286, 0.7857, 1.0000
A60	0.4286, 0.7857, 1.0000	0.5714, 0.8571, 1.0000	0.4286, 0.7500, 1.0000	0.4286, 0.8214, 1.0000	0.5714, 0.7857, 1.0000	0.4286, 0.8214, 1.0000	0.4286, 0.7143, 1.0000
A61	0.2857, 0.6786, 1.0000	0.2857, 0.6071, 0.8571	0.4286, 0.7143, 1.0000	0.4286, 0.7857, 1.0000	0.5714, 0.8214, 1.0000	0.4286, 0.7857, 1.0000	0.2857, 0.6429, 1.0000
A62	0.2857, 0.7857, 1.0000	0.2857, 0.6429, 1.0000	0.4286, 0.7500, 1.0000	0.4286, 0.8214, 1.0000	0.5714, 0.8571, 1.0000	0.4286, 0.8214, 1.0000	0.2857, 0.6786, 1.0000
A63	0.1429, 0.5357, 0.8571	0.1429, 0.5357, 0.8571	0.4286, 0.7143, 1.0000	0.5714, 0.8571, 1.0000	0.5714, 0.8571, 1.0000	0.2857, 0.6429, 1.0000	0.2857, 0.5357, 0.8571
A64	0.4286, 0.7500, 1.0000	0.2857, 0.6786, 1.0000	0.4286, 0.7500, 1.0000	0.4286, 0.7500, 1.0000	0.5714, 0.7857, 1.0000	0.4286, 0.7143, 1.0000	0.2857, 0.5714, 0.8571
A65	0.2857, 0.5714, 1.0000	0.4286, 0.7143, 1.0000	0.4286, 0.8214, 1.0000	0.5714, 0.7857, 1.0000	0.5714, 0.8214, 1.0000	0.4286, 0.7857, 1.0000	0.2857, 0.5714, 0.8571
A66	0.4286, 0.8571, 1.0000	0.4286, 0.8214, 1.0000	0.4286, 0.7500, 1.0000	0.4286, 0.8214, 1.0000	0.4286, 0.8571, 1.0000	0.2857, 0.7857, 1.0000	0.4286, 0.7857, 1.0000
A67	0.5714, 0.8571, 1.0000	0.5714, 0.8214, 1.0000	0.4286, 0.7857, 1.0000	0.4286, 0.7857, 1.0000	0.4286, 0.8214, 1.0000	0.4286, 0.7143, 1.0000	0.2857, 0.6429, 1.0000
A68	0.4286, 0.7857, 1.0000	0.4286, 0.7857, 1.0000	0.5714, 0.8214, 1.0000	0.4286, 0.7857, 1.0000	0.4286, 0.7500, 1.0000	0.4286, 0.7500, 1.0000	0.2857, 0.6071, 0.8571

At the end of these two steps, Table 3.4 is obtained with the fuzzy normalized weights for each criterion. The type represents the category of the criteria, the seven criteria are of beneficial type (+). The normalized TFN weights for each criterion will be the weights for each criterion in the fuzzy normalized decision matrix in Table 3.5.

TABLE 3.4
Properties of the Different Criteria

	Criterion	Type	TFN Weight	Weight TFN Standardized
1	C1	+	5, 6.5, 7	0.2500, 0.2321, 0.2059
2	C2	+	1, 4.25, 7	0.1477, 0.1518, 0.1471
3	C3	+	4, 5.25, 7	0.1932, 0.1875, 0.1838
4	C4	+	1, 2.5, 4	0.0795, 0.0893, 0.1029
5	C5	+	1, 2.5, 5	0.0909, 0.0893, 0.1029
6	C6	+	1, 3.75, 7	0.1364, 0.1339, 0.1397
7	C7	+	1, 3.25, 7	0.1023, 0.1161, 0.1176

The step of creating the fuzzy-weighted normalized decision matrix aims to consider the relative importance of each criterion to the normalized evaluations previously obtained. Applying the fuzzy weights of the criteria to the normalized values ensures that the most relevant criteria have the greatest impact on the decision process.

The process consists of combining the fuzzy normalized decision matrix with the vector of fuzzy weights using the formula (3.6) to obtain the fuzzy-weighted decision matrix. The obtained fuzzy-weighted matrix illustrated in Table 3.5 allows to quantitatively represent both the valuation of the alternatives and the relative importance of each criterion (Wang & Chang, 2007). This matrix constitutes the basis for the calculation of the distance to the fuzzy positive ideal fuzzy solution (FPIS) and fuzzy negative ideal fuzzy solution (FNIS), crucial steps in the Fuzzy TOPSIS methodology (Fathi et al., 2011).

$$\tilde{v} = \tilde{r}_{\text{ij ij}} \otimes \tilde{w}_{\text{j}} \tag{3.6}$$

$$\begin{aligned} \tilde{v}_{ij} &= \tilde{r}_{ij} x\ w_j = \tilde{A}_1 \otimes \tilde{A}_2 \\ &= (a_1, b_1, c_1) \otimes (a_2, b_2, c_2) \\ &= (a_1 * a_2, b_1 * b_2, c_1 * c_2) \end{aligned}$$

where:

$\tilde{r}_{\text{ij}}$: fuzzy normalized value of criterion j for alternative i.
$\tilde{w}_{\text{j}}$: fuzzy weight of criterion j.
$\tilde{v}_{\text{ij}}$: fuzzy normalized weighted fuzzy value.

TABLE 3.5
Normalized and Fuzzy-Weighted Decision Matrix

	0.25	0.232	0.206	0.148	0.152	0.147	0.193	0.188	0.184	0.08	0.089	0.103	0.091	0.089	0.103	0.136	0.134	0.14	0.102	0.116	0.118
Wj		C1			C2			C3			C4			C5			C6			C7	
A1	0.2143 ,	0.2321 ,	0.2059	0.1055 ,	0.1464 ,	0.1471	0.0552 ,	0.1339 ,	0.1838	0.0000 ,	0.0734 ,	0.1029	0.0390 ,	0.0542 ,	0.0882	0.0779 ,	0.1100 ,	0.1397	0.0438 ,	0.0954 ,	0.1176
A2	0.0714 ,	0.1326 ,	0.1765	0.0422 ,	0.0976 ,	0.1471	0.0552 ,	0.1138 ,	0.1575	0.0227 ,	0.0606 ,	0.1029	0.0130 ,	0.0478 ,	0.1029	0.0195 ,	0.0670 ,	0.1197	0.0438 ,	0.0829 ,	0.1176
A3	0.1429 ,	0.2155 ,	0.2059	0.0844 ,	0.1247 ,	0.1471	0.1104 ,	0.1339 ,	0.1575	0.0227 ,	0.0734 ,	0.1029	0.0260 ,	0.0702 ,	0.1029	0.0779 ,	0.1196 ,	0.1397	0.0585 ,	0.0954 ,	0.1176
A4	0.1429 ,	0.2072 ,	0.2059	0.0844 ,	0.1301 ,	0.1471	0.0828 ,	0.1272 ,	0.1838	0.0568 ,	0.0829 ,	0.1029	0.0519 ,	0.0765 ,	0.1029	0.0585 ,	0.1148 ,	0.1397	0.0438 ,	0.0954 ,	0.1176
A5	0.1429 ,	0.1989 ,	0.2059	0.0633 ,	0.1139 ,	0.1471	0.0828 ,	0.1272 ,	0.1838	0.0454 ,	0.0797 ,	0.1029	0.0519 ,	0.0734 ,	0.1029	0.0585 ,	0.1100 ,	0.1397	0.0438 ,	0.0954 ,	0.1176
A6	0.1429 ,	0.1907 ,	0.2059	0.0633 ,	0.0922 ,	0.1261	0.0828 ,	0.1540 ,	0.1838	0.0341 ,	0.0765 ,	0.1029	0.0390 ,	0.0670 ,	0.1029	0.0585 ,	0.0909 ,	0.1197	0.0292 ,	0.0663 ,	0.1008
A7	0.1429 ,	0.1989 ,	0.2059	0.0633 ,	0.1247 ,	0.1471	0.0552 ,	0.1138 ,	0.1838	0.0227 ,	0.0574 ,	0.1029	0.0260 ,	0.0478 ,	0.0735	0.0585 ,	0.0956 ,	0.1397	0.0585 ,	0.0954 ,	0.1176
A8	0.0714 ,	0.1989 ,	0.2059	0.0422 ,	0.1247 ,	0.1471	0.0552 ,	0.1406 ,	0.1838	0.0227 ,	0.0734 ,	0.1029	0.0260 ,	0.0638 ,	0.1029	0.0390 ,	0.1100 ,	0.1397	0.0292 ,	0.0954 ,	0.1176
A9	0.1429 ,	0.2072 ,	0.2059	0.1055 ,	0.1355 ,	0.1471	0.0828 ,	0.1272 ,	0.1838	0.0341 ,	0.0670 ,	0.1029	0.0390 ,	0.0670 ,	0.1029	0.0585 ,	0.1148 ,	0.1397	0.0292 ,	0.0871 ,	0.1176
A10	0.1071 ,	0.1989 ,	0.2059	0.0633 ,	0.1301 ,	0.1471	0.0828 ,	0.1339 ,	0.1838	0.0227 ,	0.0638 ,	0.1029	0.0260 ,	0.0638 ,	0.1029	0.0390 ,	0.1100 ,	0.1397	0.0146 ,	0.0912 ,	0.1176
A11	0.1429 ,	0.2155 ,	0.2059	0.0633 ,	0.1193 ,	0.1471	0.0828 ,	0.1540 ,	0.1838	0.0341 ,	0.0765 ,	0.1029	0.0390 ,	0.0734 ,	0.1029	0.0585 ,	0.1100 ,	0.1397	0.0438 ,	0.0829 ,	0.1176
A12	0.1071 ,	0.1907 ,	0.2059	0.0633 ,	0.1247 ,	0.1471	0.0828 ,	0.1540 ,	0.1838	0.0341 ,	0.0734 ,	0.1029	0.0390 ,	0.0702 ,	0.1029	0.0585 ,	0.1100 ,	0.1397	0.0438 ,	0.0871 ,	0.1176
A13	0.1429 ,	0.1989 ,	0.2059	0.0844 ,	0.1410 ,	0.1471	0.0552 ,	0.1205 ,	0.1575	0.0341 ,	0.0765 ,	0.1029	0.0390 ,	0.0765 ,	0.1029	0.0585 ,	0.1148 ,	0.1397	0.0292 ,	0.0954 ,	0.1176
A14	0.1429 ,	0.2072 ,	0.2059	0.0633 ,	0.1193 ,	0.1471	0.0828 ,	0.1339 ,	0.1838	0.0227 ,	0.0638 ,	0.1029	0.0130 ,	0.0638 ,	0.1029	0.0390 ,	0.1100 ,	0.1397	0.0438 ,	0.0954 ,	0.1176
A15	0.1071 ,	0.1907 ,	0.2059	0.0633 ,	0.1301 ,	0.1471	0.0552 ,	0.1272 ,	0.1838	0.0227 ,	0.0734 ,	0.1029	0.0390 ,	0.0734 ,	0.1029	0.0585 ,	0.1148 ,	0.1397	0.0438 ,	0.0995 ,	0.1176
A16	0.1071 ,	0.1658 ,	0.2059	0.0633 ,	0.1084 ,	0.1471	0.1104 ,	0.1540 ,	0.1838	0.0454 ,	0.0765 ,	0.1029	0.0519 ,	0.0829 ,	0.1029	0.0585 ,	0.1004 ,	0.1397	0.0292 ,	0.0663 ,	0.1176
A17	0.1429 ,	0.1741 ,	0.2059	0.0844 ,	0.1355 ,	0.1471	0.0828 ,	0.1473 ,	0.1838	0.0227 ,	0.0606 ,	0.1029	0.0260 ,	0.0542 ,	0.1029	0.0585 ,	0.1052 ,	0.1397	0.0438 ,	0.0995 ,	0.1176
A18	0.1071 ,	0.1824 ,	0.2059	0.0844 ,	0.1355 ,	0.1471	0.1104 ,	0.1540 ,	0.1838	0.0341 ,	0.0702 ,	0.1029	0.0260 ,	0.0606 ,	0.1029	0.0585 ,	0.1100 ,	0.1397	0.0585 ,	0.0995 ,	0.1176
A19	0.1429 ,	0.2072 ,	0.2059	0.0844 ,	0.1355 ,	0.1471	0.0828 ,	0.1473 ,	0.1838	0.0341 ,	0.0702 ,	0.1029	0.0390 ,	0.0670 ,	0.1029	0.0779 ,	0.1148 ,	0.1397	0.0585 ,	0.1037 ,	0.1176
A20	0.0714 ,	0.1741 ,	0.2059	0.0633 ,	0.1193 ,	0.1471	0.0828 ,	0.1473 ,	0.1838	0.0341 ,	0.0638 ,	0.1029	0.0260 ,	0.0542 ,	0.1029	0.0779 ,	0.1100 ,	0.1397	0.0585 ,	0.0995 ,	0.1176
A21	0.1786 ,	0.2155 ,	0.2059	0.0844 ,	0.1301 ,	0.1471	0.0828 ,	0.1406 ,	0.1838	0.0341 ,	0.0702 ,	0.1029	0.0260 ,	0.0638 ,	0.1029	0.0585 ,	0.1004 ,	0.1397	0.0438 ,	0.0871 ,	0.1176
A22	0.0714 ,	0.1658 ,	0.2059	0.0633 ,	0.1139 ,	0.1471	0.1104 ,	0.1674 ,	0.1838	0.0454 ,	0.0702 ,	0.1029	0.0390 ,	0.0670 ,	0.1029	0.0585 ,	0.1052 ,	0.1397	0.0438 ,	0.0705 ,	0.1008

(Continued)

TABLE 3.5 *(Continued)*

Normalized and Fuzzy-Weighted Decision Matrix

	0.25	0.232	0.206	0.148	0.152	0.147	0.193	0.188	0.184	0.08	0.089	0.103	0.091	0.089	0.103	0.136	0.134	0.14	0.102	0.116	0.118
Wj		C1			C2			C3			C4			C5			C6			C7	
A23	0.1071	0.1907	0.2059	0.0844	0.1193	0.1471	0.0828	0.1473	0.1838	0.0227	0.0638	0.1029	0.0390	0.0670	0.1029	0.0779	0.1148	0.1397	0.0438	0.0829	0.1176
A24	0.0357	0.1078	0.1765	0.0211	0.0759	0.1471	0.0276	0.0938	0.1575	0.0114	0.0542	0.1029	0.0390	0.0638	0.1029	0.0585	0.0765	0.0998	0.0292	0.0705	0.1008
A25	0.1429	0.2072	0.2059	0.0844	0.1247	0.1471	0.0828	0.1406	0.1838	0.0454	0.0765	0.1029	0.0390	0.0702	0.1029	0.0585	0.0956	0.1397	0.0292	0.0622	0.1008
A26	0.0714	0.1492	0.2059	0.0633	0.0976	0.1261	0.1104	0.1540	0.1838	0.0341	0.0734	0.1029	0.0519	0.0765	0.1029	0.0585	0.1148	0.1397	0.0292	0.0912	0.1176
A27	0.1429	0.2155	0.2059	0.0633	0.1247	0.1471	0.0828	0.1138	0.1575	0.0227	0.0510	0.0882	0.0260	0.0510	0.0882	0.0390	0.1004	0.1397	0.0438	0.0788	0.1176
A28	0.1429	0.1989	0.2059	0.0633	0.1247	0.1471	0.0828	0.1406	0.1838	0.0227	0.0670	0.1029	0.0260	0.0638	0.1029	0.0390	0.1052	0.1397	0.0292	0.0912	0.1176
A29	0.1429	0.1989	0.2059	0.0633	0.1193	0.1471	0.0828	0.1473	0.1838	0.0454	0.0702	0.1029	0.0390	0.0702	0.1029	0.0585	0.1100	0.1397	0.0438	0.0954	0.1176
A30	0.1429	0.2072	0.2059	0.0422	0.1193	0.1471	0.0828	0.1406	0.1838	0.0341	0.0734	0.1029	0.0260	0.0702	0.1029	0.0585	0.1100	0.1397	0.0292	0.0912	0.1176
A31	0.1071	0.1824	0.2059	0.0422	0.1030	0.1471	0.0828	0.1406	0.1838	0.0227	0.0606	0.1029	0.0260	0.0510	0.0882	0.0390	0.0813	0.1197	0.0438	0.0871	0.1176
A32	0.1071	0.1824	0.2059	0.0844	0.1355	0.1471	0.0552	0.1138	0.1575	0.0227	0.0574	0.1029	0.0260	0.0510	0.0882	0.0390	0.1052	0.1397	0.0292	0.0871	0.1176
A33	0.0714	0.1326	0.1765	0.0633	0.1139	0.1471	0.0828	0.1406	0.1838	0.0341	0.0606	0.1029	0.0390	0.0638	0.1029	0.0585	0.1052	0.1397	0.0438	0.0954	0.1176
A34	0.0357	0.1409	0.2059	0.0633	0.1030	0.1471	0.1104	0.1674	0.1838	0.0341	0.0638	0.1029	0.0390	0.0670	0.1029	0.0585	0.0956	0.1397	0.0438	0.0788	0.1176
A35	0.0714	0.1326	0.2059	0.0422	0.0976	0.1471	0.1104	0.1607	0.1838	0.0341	0.0638	0.1029	0.0260	0.0606	0.1029	0.0779	0.1004	0.1397	0.0292	0.0622	0.1008
A36	0.0714	0.1326	0.2059	0.0633	0.1030	0.1471	0.1104	0.1607	0.1838	0.0341	0.0638	0.1029	0.0390	0.0638	0.1029	0.0585	0.0956	0.1397	0.0292	0.0705	0.1008
A37	0.0714	0.1326	0.2059	0.0633	0.1030	0.1471	0.1104	0.1674	0.1838	0.0227	0.0574	0.1029	0.0390	0.0638	0.1029	0.0390	0.0956	0.1397	0.0292	0.0622	0.1008
A38	0.0714	0.1575	0.2059	0.0633	0.1139	0.1471	0.1104	0.1607	0.1838	0.0341	0.0542	0.0882	0.0390	0.0606	0.1029	0.0585	0.1004	0.1397	0.0438	0.0788	0.1176
A39	0.0714	0.1409	0.2059	0.0422	0.0922	0.1471	0.0552	0.1406	0.1838	0.0341	0.0702	0.1029	0.0390	0.0734	0.1029	0.0585	0.1004	0.1397	0.0292	0.0622	0.1008
A40	0.1071	0.1741	0.2059	0.0422	0.1139	0.1471	0.0828	0.1272	0.1838	0.0227	0.0542	0.0882	0.0260	0.0510	0.0882	0.0390	0.1052	0.1397	0.0438	0.0954	0.1176
A41	0.1071	0.1658	0.2059	0.0422	0.1139	0.1471	0.0828	0.1272	0.1838	0.0227	0.0542	0.0882	0.0260	0.0510	0.0882	0.0390	0.1052	0.1397	0.0438	0.0954	0.1176
A42	0.1071	0.1907	0.2059	0.0633	0.1247	0.1471	0.1104	0.1607	0.1838	0.0227	0.0702	0.1029	0.0260	0.0702	0.1029	0.0585	0.1100	0.1397	0.0438	0.0954	0.1176
A43	0.1071	0.1907	0.2059	0.0844	0.1247	0.1471	0.1104	0.1607	0.1838	0.0227	0.0574	0.1029	0.0260	0.0574	0.1029	0.0585	0.1052	0.1397	0.0438	0.0954	0.1176
A44	0.1071	0.1989	0.2059	0.0633	0.1193	0.1471	0.0552	0.1473	0.1838	0.0227	0.0574	0.1029	0.0260	0.0542	0.0882	0.0585	0.1100	0.1397	0.0585	0.0995	0.1176

A45	0.0714 , 0.1907 , 0.2059	0.0633 , 0.1084 , 0.1471	0.1380 , 0.1741 , 0.1838	0.0227 , 0.0638 , 0.1029	0.0390 , 0.0702 , 0.1029	0.0585 , 0.1052 , 0.1397	0.0292 , 0.0871 , 0.1176
A46	0.0714 , 0.1824 , 0.2059	0.0633 , 0.1139 , 0.1471	0.0828 , 0.1473 , 0.1838	0.0227 , 0.0638 , 0.1029	0.0390 , 0.0702 , 0.1029	0.0585 , 0.1004 , 0.1397	0.0438 , 0.0746 , 0.1008
A47	0.0714 , 0.1409 , 0.2059	0.0633 , 0.1030 , 0.1471	0.1104 , 0.1674 , 0.1838	0.0227 , 0.0606 , 0.1029	0.0260 , 0.0670 , 0.1029	0.0390 , 0.0956 , 0.1397	0.0292 , 0.0705 , 0.1008
A48	0.1071 , 0.1492 , 0.2059	0.0844 , 0.1084 , 0.1261	0.1104 , 0.1540 , 0.1838	0.0227 , 0.0606 , 0.1029	0.0260 , 0.0606 , 0.1029	0.0585 , 0.1004 , 0.1397	0.0438 , 0.0788 , 0.1176
A49	0.0714 , 0.1492 , 0.2059	0.0633 , 0.1247 , 0.1471	0.1104 , 0.1607 , 0.1838	0.0227 , 0.0702 , 0.1029	0.0390 , 0.0734 , 0.1029	0.0390 , 0.1052 , 0.1397	0.0292 , 0.0746 , 0.1176
A50	0.0714 , 0.1658 , 0.2059	0.0844 , 0.1301 , 0.1471	0.0552 , 0.1071 , 0.1575	0.0227 , 0.0702 , 0.1029	0.0260 , 0.0510 , 0.0882	0.0585 , 0.1100 , 0.1397	0.0438 , 0.0954 , 0.1176
A51	0.1429 , 0.1989 , 0.2059	0.0844 , 0.1301 , 0.1471	0.0828 , 0.1138 , 0.1575	0.0341 , 0.0734 , 0.1029	0.0390 , 0.0542 , 0.0882	0.0779 , 0.1148 , 0.1397	0.0292 , 0.0912 , 0.1176
A52	0.1429 , 0.2072 , 0.2059	0.0633 , 0.1193 , 0.1471	0.0552 , 0.1138 , 0.1575	0.0454 , 0.0765 , 0.1029	0.0390 , 0.0606 , 0.1029	0.0585 , 0.1100 , 0.1397	0.0438 , 0.0954 , 0.1176
A53	0.1429 , 0.2155 , 0.2059	0.0844 , 0.1301 , 0.1471	0.0828 , 0.1473 , 0.1838	0.0341 , 0.0734 , 0.1029	0.0390 , 0.0702 , 0.1029	0.0779 , 0.1148 , 0.1397	0.0585 , 0.0995 , 0.1176
A54	0.1429 , 0.2072 , 0.2059	0.0844 , 0.1301 , 0.1471	0.0828 , 0.1339 , 0.1838	0.0454 , 0.0765 , 0.1029	0.0260 , 0.0702 , 0.1029	0.0585 , 0.1100 , 0.1397	0.0438 , 0.0954 , 0.1176
A55	0.1786 , 0.2155 , 0.2059	0.0633 , 0.1247 , 0.1471	0.1104 , 0.1406 , 0.1838	0.0341 , 0.0734 , 0.1029	0.0390 , 0.0670 , 0.1029	0.0585 , 0.1100 , 0.1397	0.0292 , 0.0788 , 0.1176
A56	0.1071 , 0.1907 , 0.2059	0.0633 , 0.1247 , 0.1471	0.0828 , 0.1339 , 0.1838	0.0341 , 0.0702 , 0.1029	0.0390 , 0.0702 , 0.1029	0.0585 , 0.1100 , 0.1397	0.0438 , 0.0954 , 0.1176
A57	0.1786 , 0.2155 , 0.2059	0.0633 , 0.1247 , 0.1471	0.0552 , 0.1205 , 0.1575	0.0454 , 0.0670 , 0.1029	0.0519 , 0.0702 , 0.1029	0.0585 , 0.1052 , 0.1397	0.0585 , 0.0995 , 0.1176
A58	0.1786 , 0.2072 , 0.2059	0.0633 , 0.1193 , 0.1471	0.0828 , 0.1138 , 0.1575	0.0454 , 0.0734 , 0.1029	0.0390 , 0.0670 , 0.1029	0.0585 , 0.1052 , 0.1397	0.0438 , 0.0912 , 0.1176
A59	0.1429 , 0.1989 , 0.2059	0.0844 , 0.1301 , 0.1471	0.1104 , 0.1473 , 0.1838	0.0341 , 0.0638 , 0.1029	0.0390 , 0.0574 , 0.0882	0.0585 , 0.1100 , 0.1397	0.0438 , 0.0912 , 0.1176
A60	0.1071 , 0.1824 , 0.2059	0.0844 , 0.1301 , 0.1471	0.0828 , 0.1406 , 0.1838	0.0341 , 0.0734 , 0.1029	0.0519 , 0.0702 , 0.1029	0.0585 , 0.1100 , 0.1397	0.0438 , 0.0829 , 0.1176
A61	0.0714 , 0.1575 , 0.2059	0.0422 , 0.0922 , 0.1261	0.0828 , 0.1339 , 0.1838	0.0341 , 0.0702 , 0.1029	0.0519 , 0.0734 , 0.1029	0.0585 , 0.1052 , 0.1397	0.0292 , 0.0746 , 0.1176
A62	0.0714 , 0.1824 , 0.2059	0.0422 , 0.0976 , 0.1471	0.0828 , 0.1406 , 0.1838	0.0341 , 0.0734 , 0.1029	0.0519 , 0.0765 , 0.1029	0.0585 , 0.1100 , 0.1397	0.0292 , 0.0788 , 0.1176
A63	0.0357 , 0.1243 , 0.1765	0.0211 , 0.0813 , 0.1261	0.0828 , 0.1339 , 0.1838	0.0454 , 0.0765 , 0.1029	0.0519 , 0.0765 , 0.1029	0.0390 , 0.0861 , 0.1397	0.0292 , 0.0622 , 0.1008
A64	0.1071 , 0.1741 , 0.2059	0.0422 , 0.1030 , 0.1471	0.0828 , 0.1406 , 0.1838	0.0341 , 0.0670 , 0.1029	0.0519 , 0.0702 , 0.1029	0.0585 , 0.0956 , 0.1397	0.0292 , 0.0663 , 0.1008
A65	0.0714 , 0.1326 , 0.2059	0.0633 , 0.1084 , 0.1471	0.0828 , 0.1540 , 0.1838	0.0454 , 0.0702 , 0.1029	0.0519 , 0.0734 , 0.1029	0.0585 , 0.1052 , 0.1397	0.0292 , 0.0663 , 0.1008
A66	0.1071 , 0.1989 , 0.2059	0.0633 , 0.1247 , 0.1471	0.0828 , 0.1406 , 0.1838	0.0341 , 0.0734 , 0.1029	0.0390 , 0.0765 , 0.1029	0.0390 , 0.1052 , 0.1397	0.0438 , 0.0912 , 0.1176
A67	0.1429 , 0.1989 , 0.2059	0.0844 , 0.1247 , 0.1471	0.0828 , 0.1473 , 0.1838	0.0341 , 0.0702 , 0.1029	0.0390 , 0.0734 , 0.1029	0.0585 , 0.0956 , 0.1397	0.0292 , 0.0746 , 0.1176
A68	0.1071 , 0.1824 , 0.2059	0.0633 , 0.1193 , 0.1471	0.1104 , 0.1540 , 0.1838	0.0341 , 0.0702 , 0.1029	0.0390 , 0.0670 , 0.1029	0.0585 , 0.1004 , 0.1397	0.0292 , 0.0705 , 0.1008

Step 4: Determination of the fuzzy positive ideal solution (FPIS) and the fuzzy negative ideal solution (FNIS).

In this step, FPIS and FNIS are calculated for each alternative. The FPIS and FNIS solutions are key to the subsequent steps. The formulas used are defined as follows:

To calculate FPIS the following formula was used:

$$A^+ = \left(\tilde{v}_1^+, \tilde{v}_2^+, \ldots, \tilde{v}_n^+\right) \text{ where } \tilde{v}_j^+ = \max_i (v_{ij}) \tag{3.7}$$

where:

$\tilde{v}_j^+$ represents the best possible value of criterion j.

The following formula was used to calculate FNIS:

$$A^- = \left(\tilde{v}_1^-, \tilde{v}_2^-, \ldots, \tilde{v}_n^-\right) \text{ where } \tilde{v}_j^- = \min_i (v_{ij}) \tag{3.8}$$

where:

$\tilde{v}_j^-$ represents the worst possible value of criterion j, taking into account the weighted fuzzy values of all alternatives.

$\max_{(i)}$, $\min_i$: operators that take the maximum or minimum value among all alternatives i for each component of the TFN.

Table 3.6 shows the result of the calculation of the FPIS and the FNIS.

TABLE 3.6
Positive and Negative Ideal Solutions

	Ideal Positive (FPIS)	Ideal Negative (FNIS)
C1	(0.2143, 0.2321, 0.2059)	(0.0357, 0.1078, 0.1765)
C2	(0.1055, 0.1464, 0.1471)	(0.0211, 0.0759, 0.1261)
C3	(0.1380, 0.1741, 0.1838)	(0.0276, 0.0938, 0.1575)
C4	(0.0568, 0.0829, 0.1029)	(0.0000, 0.0510, 0.0882)
C5	(0.0519, 0.0829, 0.1029)	(0.0130, 0.0478, 0.0735)
C6	(0.0779, 0.1196, 0.1397)	(0.0195, 0.0670, 0.0998)
C7	(0.0585, 0.1037, 0.1176)	(0.0146, 0.0622, 0.1008)

FPIS and FNIS are the basis for determining the *closeness coefficient (CC)* that indicates how favorable or good each alternative is relative to the ideal (Roshandel et al., 2021).

Step 5: Calculate the distance of each alternative with respect to the FPIS and FNIS.

Euclidean distance is used as a metric to quantify the degree of proximity of each alternative to the established ideal points. The formula (3.9) was used in this step to calculate the distance of each alternative with respect to the FPIS and the FNIS.

$$d(\tilde{x},\tilde{y}) = \sqrt{\frac{1}{3}\left[(a_1 - a_2)^2 + (b_1 - b_2)^2 + (c_1 - c_2)^2\right]} \tag{3.9}$$

This step allows to transform the fuzzy judgments into a numerical metric reflecting the relative closeness to the ideal (closeness coefficient). It is the connection between the fuzzy judgments with an objective and measurable ranking. The distance of alternatives from FPIS is illustrated in Table 3.7 and the distance of alternatives from FNIS is illustrated in Table 3.8.

TABLE 3.7
Distance of Alternatives from the FPIS

	C1	C2	C3	C4	C5	C6	C7
A1	0.0000	0.0000	0.0531	0.0332	0.0201	0.0055	0.0097
A2	0.1019	0.0461	0.0610	0.0235	0.0303	0.0468	0.0146
A3	0.0423	0.0175	0.0320	0.0204	0.0167	0.0000	0.0048
A4	0.0437	0.0154	0.0418	0.0000	0.0037	0.0116	0.0097
A5	0.0455	0.0308	0.0418	0.0068	0.0055	0.0125	0.0097
A6	0.0477	0.0415	0.0339	0.0136	0.0119	0.0231	0.0290
A7	0.0455	0.0274	0.0591	0.0246	0.0304	0.0178	0.0048
A8	0.0847	0.0386	0.0516	0.0204	0.0186	0.0232	0.0175
A9	0.0437	0.0063	0.0418	0.0160	0.0119	0.0116	0.0194
A10	0.0648	0.0261	0.0394	0.0226	0.0186	0.0232	0.0263
A11	0.0423	0.0290	0.0339	0.0136	0.0093	0.0125	0.0146
A12	0.0663	0.0274	0.0339	0.0142	0.0105	0.0125	0.0128
A13	0.0455	0.0126	0.0589	0.0136	0.0084	0.0116	0.0175
A14	0.0437	0.0290	0.0394	0.0226	0.0251	0.0232	0.0097
A15	0.0663	0.0261	0.0549	0.0204	0.0093	0.0116	0.0088
A16	0.0727	0.0328	0.0197	0.0075	0.0000	0.0158	0.0274
A17	0.0531	0.0137	0.0354	0.0235	0.0223	0.0140	0.0088
A18	0.0682	0.0137	0.0197	0.0150	0.0198	0.0125	0.0024
A19	0.0437	0.0137	0.0354	0.0150	0.0119	0.0028	0.0000
A20	0.0890	0.0290	0.0354	0.0171	0.0223	0.0055	0.0024
A21	0.0227	0.0154	0.0373	0.0150	0.0186	0.0158	0.0128
A22	0.0909	0.0308	0.0164	0.0099	0.0119	0.0140	0.0231
A23	0.0663	0.0198	0.0354	0.0226	0.0119	0.0028	0.0146
A24	0.1268	0.0635	0.0803	0.0310	0.0134	0.0357	0.0273
A25	0.0437	0.0175	0.0373	0.0075	0.0105	0.0178	0.0309

(Continued)

TABLE 3.7 ***(Continued)***

Distance of Alternatives from the FPIS

	C1	C2	C3	C4	C5	C6	C7
A26	0.0954	0.0392	0.0197	0.0142	0.0037	0.0116	0.0183
A27	0.0423	0.0274	0.0496	0.0282	0.0252	0.0251	0.0167
A28	0.0455	0.0274	0.0373	0.0217	0.0186	0.0240	0.0183
A29	0.0455	0.0290	0.0354	0.0099	0.0105	0.0125	0.0097
A30	0.0437	0.0398	0.0373	0.0142	0.0167	0.0125	0.0183
A31	0.0682	0.0443	0.0373	0.0235	0.0252	0.0336	0.0128
A32	0.0682	0.0137	0.0610	0.0246	0.0252	0.0240	0.0194
A33	0.1019	0.0308	0.0373	0.0184	0.0134	0.0140	0.0097
A34	0.1158	0.0349	0.0164	0.0171	0.0119	0.0178	0.0167
A35	0.1005	0.0461	0.0177	0.0171	0.0198	0.0110	0.0309
A36	0.1005	0.0349	0.0177	0.0171	0.0134	0.0178	0.0273
A37	0.1005	0.0349	0.0164	0.0246	0.0134	0.0264	0.0309
A38	0.0930	0.0308	0.0177	0.0228	0.0149	0.0158	0.0167
A39	0.0978	0.0481	0.0516	0.0150	0.0093	0.0158	0.0309
A40	0.0703	0.0411	0.0418	0.0271	0.0252	0.0240	0.0097
A41	0.0727	0.0411	0.0418	0.0271	0.0252	0.0240	0.0097
A42	0.0663	0.0274	0.0177	0.0210	0.0167	0.0125	0.0097
A43	0.0663	0.0175	0.0177	0.0246	0.0210	0.0140	0.0097
A44	0.0648	0.0290	0.0502	0.0246	0.0239	0.0125	0.0024
A45	0.0859	0.0328	0.0000	0.0226	0.0105	0.0140	0.0194
A46	0.0873	0.0308	0.0354	0.0226	0.0105	0.0158	0.0211
A47	0.0978	0.0349	0.0164	0.0235	0.0176	0.0264	0.0273
A48	0.0782	0.0279	0.0197	0.0235	0.0198	0.0158	0.0167
A49	0.0954	0.0274	0.0177	0.0210	0.0093	0.0240	0.0238
A50	0.0909	0.0154	0.0633	0.0210	0.0252	0.0125	0.0097
A51	0.0455	0.0154	0.0496	0.0142	0.0201	0.0028	0.0183
A52	0.0437	0.0290	0.0610	0.0075	0.0149	0.0125	0.0097
A53	0.0423	0.0154	0.0354	0.0142	0.0105	0.0028	0.0024
A54	0.0437	0.0154	0.0394	0.0075	0.0167	0.0125	0.0097
A55	0.0227	0.0274	0.0251	0.0142	0.0119	0.0125	0.0222
A56	0.0663	0.0274	0.0394	0.0150	0.0105	0.0125	0.0097
A57	0.0227	0.0274	0.0589	0.0113	0.0074	0.0140	0.0024
A58	0.0251	0.0290	0.0496	0.0086	0.0119	0.0140	0.0111
A59	0.0455	0.0154	0.0222	0.0171	0.0186	0.0125	0.0111
A60	0.0682	0.0154	0.0373	0.0142	0.0074	0.0125	0.0146
A61	0.0930	0.0496	0.0394	0.0150	0.0055	0.0140	0.0238
A62	0.0873	0.0461	0.0373	0.0142	0.0037	0.0125	0.0222
A63	0.1216	0.0627	0.0394	0.0075	0.0037	0.0297	0.0309
A64	0.0703	0.0443	0.0373	0.0160	0.0074	0.0178	0.0290
A65	0.1005	0.0328	0.0339	0.0099	0.0055	0.0140	0.0290
A66	0.0648	0.0274	0.0373	0.0142	0.0084	0.0240	0.0111
A67	0.0455	0.0175	0.0354	0.0150	0.0093	0.0178	0.0238
A68	0.0682	0.0290	0.0197	0.0150	0.0119	0.0158	0.0273

TABLE 3.8
Distance of Alternatives from FNIS

	C1	C2	C3	C4	C5	C6	C7
A1	0.1268	0.0646	0.0320	0.0154	0.0176	0.0478	0.0273
A2	0.0251	0.0213	0.0197	0.0166	0.0170	0.0115	0.0229
A3	0.0894	0.0477	0.0531	0.0203	0.0226	0.0509	0.0332
A4	0.0861	0.0496	0.0402	0.0385	0.0327	0.0424	0.0273
A5	0.0830	0.0349	0.0402	0.0322	0.0318	0.0407	0.0273
A6	0.0800	0.0261	0.0496	0.0260	0.0252	0.0288	0.0088
A7	0.0830	0.0392	0.0249	0.0160	0.0075	0.0362	0.0332
A8	0.0590	0.0330	0.0349	0.0203	0.0207	0.0357	0.0231
A9	0.0861	0.0609	0.0402	0.0233	0.0252	0.0424	0.0193
A10	0.0690	0.0415	0.0422	0.0173	0.0207	0.0357	0.0194
A11	0.0894	0.0370	0.0496	0.0260	0.0270	0.0407	0.0229
A12	0.0654	0.0392	0.0496	0.0250	0.0261	0.0407	0.0242
A13	0.0830	0.0538	0.0222	0.0260	0.0281	0.0424	0.0231
A14	0.0861	0.0370	0.0422	0.0173	0.0193	0.0357	0.0273
A15	0.0654	0.0415	0.0293	0.0203	0.0270	0.0424	0.0290
A16	0.0558	0.0331	0.0610	0.0313	0.0347	0.0376	0.0131
A17	0.0747	0.0517	0.0469	0.0166	0.0189	0.0391	0.0290
A18	0.0620	0.0517	0.0610	0.0241	0.0200	0.0407	0.0346
A19	0.0861	0.0517	0.0469	0.0241	0.0252	0.0493	0.0362
A20	0.0467	0.0370	0.0469	0.0227	0.0189	0.0478	0.0346
A21	0.1047	0.0496	0.0445	0.0241	0.0207	0.0376	0.0242
A22	0.0428	0.0349	0.0658	0.0297	0.0252	0.0391	0.0175
A23	0.0654	0.0459	0.0469	0.0173	0.0252	0.0493	0.0229
A24	0.0000	0.0121	0.0000	0.0109	0.0244	0.0232	0.0097
A25	0.0861	0.0477	0.0445	0.0313	0.0261	0.0362	0.0084
A26	0.0359	0.0274	0.0610	0.0250	0.0327	0.0424	0.0211
A27	0.0894	0.0392	0.0339	0.0131	0.0115	0.0321	0.0217
A28	0.0830	0.0392	0.0445	0.0181	0.0207	0.0338	0.0211
A29	0.0830	0.0370	0.0469	0.0297	0.0261	0.0407	0.0273
A30	0.0861	0.0304	0.0445	0.0250	0.0226	0.0407	0.0211
A31	0.0620	0.0232	0.0445	0.0166	0.0115	0.0181	0.0242
A32	0.0620	0.0517	0.0197	0.0160	0.0115	0.0338	0.0193
A33	0.0251	0.0349	0.0445	0.0221	0.0244	0.0391	0.0273
A34	0.0256	0.0314	0.0658	0.0227	0.0252	0.0362	0.0217
A35	0.0303	0.0213	0.0633	0.0227	0.0200	0.0452	0.0084
A36	0.0303	0.0314	0.0633	0.0227	0.0244	0.0362	0.0097
A37	0.0303	0.0314	0.0658	0.0160	0.0244	0.0305	0.0084
A38	0.0392	0.0349	0.0633	0.0198	0.0238	0.0376	0.0217
A39	0.0329	0.0196	0.0349	0.0241	0.0270	0.0376	0.0084
A40	0.0588	0.0279	0.0402	0.0132	0.0115	0.0338	0.0273
A41	0.0558	0.0279	0.0402	0.0132	0.0115	0.0338	0.0273
A42	0.0654	0.0392	0.0633	0.0191	0.0226	0.0407	0.0273

(Continued)

TABLE 3.8 ***(Continued)***
Distance of Alternatives from FNIS

	C1	C2	C3	C4	C5	C6	C7
A43	0.0654	0.0477	0.0633	0.0160	0.0194	0.0391	0.0273
A44	0.0690	0.0370	0.0380	0.0160	0.0119	0.0407	0.0346
A45	0.0548	0.0331	0.0803	0.0173	0.0261	0.0391	0.0193
A46	0.0507	0.0349	0.0469	0.0173	0.0261	0.0376	0.0183
A47	0.0329	0.0314	0.0658	0.0166	0.0216	0.0305	0.0097
A48	0.0506	0.0411	0.0610	0.0166	0.0200	0.0376	0.0217
A49	0.0359	0.0392	0.0633	0.0191	0.0270	0.0338	0.0147
A50	0.0428	0.0496	0.0177	0.0191	0.0115	0.0407	0.0273
A51	0.0830	0.0496	0.0339	0.0250	0.0176	0.0493	0.0211
A52	0.0861	0.0370	0.0197	0.0313	0.0238	0.0407	0.0273
A53	0.0894	0.0496	0.0469	0.0250	0.0261	0.0493	0.0346
A54	0.0861	0.0496	0.0422	0.0313	0.0226	0.0407	0.0273
A55	0.1047	0.0392	0.0570	0.0250	0.0252	0.0407	0.0160
A56	0.0654	0.0392	0.0422	0.0241	0.0261	0.0407	0.0273
A57	0.1047	0.0392	0.0222	0.0291	0.0310	0.0391	0.0346
A58	0.1019	0.0370	0.0339	0.0304	0.0252	0.0391	0.0257
A59	0.0830	0.0496	0.0589	0.0227	0.0181	0.0407	0.0257
A60	0.0620	0.0496	0.0445	0.0250	0.0310	0.0407	0.0229
A61	0.0392	0.0154	0.0422	0.0241	0.0318	0.0391	0.0147
A62	0.0507	0.0213	0.0445	0.0250	0.0327	0.0407	0.0160
A63	0.0096	0.0031	0.0422	0.0313	0.0327	0.0279	0.0084
A64	0.0588	0.0232	0.0445	0.0233	0.0310	0.0362	0.0088
A65	0.0303	0.0331	0.0496	0.0297	0.0318	0.0391	0.0088
A66	0.0690	0.0392	0.0445	0.0250	0.0281	0.0338	0.0257
A67	0.0830	0.0477	0.0469	0.0241	0.0270	0.0362	0.0147
A68	0.0620	0.0370	0.0610	0.0241	0.0252	0.0376	0.0097

Step 6: Calculate the closeness coefficient for each alternative and rank the alternatives.

To calculate the closeness coefficient, the sum of the distances of each alternative with respect to the FPIS and the FNIS obtained in Step 5 were added, using formulas (3.10) and (3.11), respectively.

With respect to FPIS:

$$d_i^* = \sum_{j=1}^{n} d\left(\tilde{v}_{ij}, \tilde{v}_j^*\right) \tag{3.10}$$

With respect to FNIS:

$$d_i^- = \sum_{j=1}^{n} d\left(\tilde{v}_{ij}, \tilde{v}_j^-\right) \tag{3.11}$$

In the Fuzzy TOPSIS method this calculation is the previous step to obtain the closeness coefficient (C_i) for each alternative and the hierarchy of these alternatives. The í closeness coefficient is calculated using the formula (3.12) as shown below:

$$CC_i = \frac{d_i^-}{d_i^- + d_i^*} \tag{3.12}$$

where:

CC_i corresponds to the closeness coefficient for *i*.
d_i^* is the distance to the positive ideal solution.
d_i^- is the distance to the negative ideal solution.

The closeness index varies between 0 and 1; the closer to 1 it is, the better the alternative (Kumar & Pamucar, 2025). The value of the closeness coefficient helps us to determine the best alternative, or in other situations Fuzzy TOPSIS is used to rank the alternatives as is our case. Alternatives closer to 1 are alternatives closer to the positive ideal (preferable) and the closest to 0 is an alternative closer to the negative ideal (less preferable). Based on the C_i values, a *ranking or score of alternatives* is established, ordered from closest to least close to the ideal. The alternative with the highest C_i is considered *the best* option, and the others are ordered accordingly.

This step concludes by simplifying all the information obtained in the previous steps and provides a frame of reference for making an informed, quantitative decision. Table 3.9 presents the calculations performed in Step 6.

TABLE 3.9
Closeness Coefficient for Each Alternative and Hierarchy of Alternatives

	Sumatoria Distancia FPIS	Sumatoria Distancia FNIS	Coeficiente de Cercanía	JERARQUÍA
A1	0.1217	0.3316	0.7315	1
A2	0.3244	0.1340	0.2924	67
A3	0.1337	0.3172	0.7035	5
A4	0.1258	0.3169	0.7158	4
A5	0.1526	0.2901	0.6553	15
A6	0.2007	0.2445	0.5492	40
A7	0.2095	0.2400	0.5339	43
A8	0.2546	0.2267	0.4709	56
A9	0.1506	0.2974	0.6638	11
A10	0.2209	0.2458	0.5266	46
A11	0.1553	0.2925	0.6531	16

(Continued)

TABLE 3.9 ***(Continued)***
Closeness Coefficient for Each Alternative and Hierarchy of Alternatives

	Sumatoria Distancia FPIS	Sumatoria Distancia FNIS	Coeficiente de Cercanía	JERARQUÍA
A12	0.1777	0.2701	0.6032	26
A13	0.1681	0.2785	0.6237	20
A14	0.1925	0.2649	0.5791	33
A15	0.1975	0.2549	0.5635	37
A16	0.1759	0.2665	0.6024	27
A17	0.1709	0.2769	0.6184	23
A18	0.1513	0.2941	0.6602	13
A19	0.1225	0.3195	0.7229	2
A20	0.2008	0.2546	0.5591	38
A21	0.1376	0.3054	0.6894	7
A22	0.1969	0.2550	0.5644	36
A23	0.1734	0.2729	0.6115	25
A24	0.3779	0.0803	0.1753	68
A25	0.1651	0.2803	0.6293	18
A26	0.2021	0.2455	0.5485	41
A27	0.2145	0.2408	0.5289	45
A28	0.1928	0.2604	0.5746	35
A29	0.1525	0.2906	0.6559	14
A30	0.1825	0.2703	0.5970	29
A31	0.2448	0.2001	0.4497	64
A32	0.2361	0.2140	0.4755	55
A33	0.2254	0.2175	0.4911	53
A34	0.2306	0.2285	0.4977	51
A35	0.2432	0.2112	0.4648	59
A36	0.2288	0.2181	0.4880	54
A37	0.2470	0.2069	0.4559	63
A38	0.2116	0.2403	0.5317	44
A39	0.2685	0.1844	0.4072	65
A40	0.2392	0.2127	0.4707	57
A41	0.2416	0.2097	0.4647	60
A42	0.1714	0.2776	0.6183	24
A43	0.1708	0.2782	0.6197	21
A44	0.2074	0.2472	0.5438	42
A45	0.1851	0.2698	0.5931	31
A46	0.2235	0.2318	0.5091	48
A47	0.2440	0.2084	0.4607	62
A48	0.2015	0.2485	0.5523	39
A49	0.2185	0.2331	0.5161	47
A50	0.2381	0.2088	0.4672	58
A51	0.1658	0.2796	0.6277	19
A52	0.1783	0.2659	0.5985	28

(Continued)

TABLE 3.9 ***(Continued)***
Closeness Coefficient for Each Alternative and Hierarchy of Alternatives

	Sumatoria Distancia FPIS	Sumatoria Distancia FNIS	Coeficiente de Cercanía	JERARQUÍA
A53	0.1230	0.3209	0.7229	3
A54	0.1449	0.2998	0.6741	10
A55	0.1360	0.3078	0.6936	6
A56	0.1809	0.2650	0.5942	30
A57	0.1441	0.2998	0.6754	9
A58	0.1491	0.2932	0.6628	12
A59	0.1424	0.2986	0.6771	8
A60	0.1696	0.2756	0.6190	22
A61	0.2404	0.2065	0.4621	61
A62	0.2234	0.2308	0.5082	49
A63	0.2955	0.1552	0.3444	66
A64	0.2222	0.2258	0.5041	50
A65	0.2256	0.2223	0.4963	52
A66	0.1871	0.2652	0.5864	32
A67	0.1643	0.2797	0.6299	17
A68	0.1869	0.2566	0.5786	34

3.3.3 Principal Component Analysis (PCA)

Crucial to the study was the use of PCA, the objective of using it was to identify the components that empirically group the items of the instrument designed to assess the efficiency of the implementation of the GTT and to identify each component that contributes to the efficient implementation of the GTT. PCA was useful to represent the principal components as a proposed alternative to the seven key areas of the ITG according to the OECD. The PCA approach maximizes the variance explained by each component, which facilitates the interpretation of the information (Mooi et al., 2018).

The six methodological steps applied in the PCA are mentioned below (George & Mallery, 2003; Loewen & Gonulal, 2015; Phakiti, 2015; Mooi et al., 2018).

Step 1: Preparation of the data matrix.

The data matrix was organized where the rows represented the evaluators' observations for each of the seven ITG criteria (completing a total of 28 rows) and in the columns the ITG-related items (68 items). The items were coded with the prefixes E, P, and M followed by the corresponding number, in order to identify to which governance mechanism the item belonged (Structure, Processes, Relational Mechanisms). Table 3.10 shows the data matrix.

TABLE 3.10
Data Matrix. Part 1

	Estructura																		
Criterio	**E01**	**E02**	**E03**	**E04**	**E05**	**E06**	**E07**	**E08**	**E09**	**E10**	**E11**	**E12**	**E13**	**E14**	**E15**	**E16**	**E17**	**E18**	**E19**
C1	7	3	7	7	7	6	7	7	7	7	7	7	7	6	7	4	6	6	7
C2	7	3	6	6	6	4	6	7	7	7	7	7	7	7	7	4	7	7	7
C3	5	4	5	4	4	6	3	5	4	6	6	7	5	4	4	5	4	5	5
C4	7	4	7	7	7	7	3	7	4	4	7	5	7	4	7	6	6	6	5
C5	4	2	7	6	6	5	4	4	4	4	7	5	7	4	6	7	4	4	5
C6	7	4	7	7	7	5	5	7	7	7	7	7	7	7	7	4	7	7	7
C7	7	5	6	7	7	4	7	7	7	7	7	7	7	7	7	4	7	7	7
C1	7	5	7	5	5	6	5	7	7	7	7	5	7	7	6	7	5	7	7
C2	7	5	5	5	5	5	6	6	6	6	6	6	7	6	6	6	6	6	6
C3	6	5	5	5	6	6	6	6	6	6	6	7	5	7	6	6	6	6	6
C4	5	5	6	6	6	6	6	6	6	6	6	7	6	6	6	6	6	5	6
C5	5	2	5	5	5	5	4	6	6	6	5	6	6	7	6	7	6	5	5
C6	5	2	6	6	5	5	4	6	6	6	5	6	6	6	6	6	6	5	5
C7	6	6	6	5	5	5	5	6	5	6	5	6	6	7	6	6	6	5	6
C1	7	4	7	7	7	6	7	7	5	6	7	7	5	7	6	5	5	5	6
C2	7	6	7	7	6	4	7	7	6	7	4	6	7	4	7	6	7	7	7
C3	3	5	5	6	5	7	4	7	5	4	7	5	5	5	6	7	7	7	7
C4	7	7	7	7	7	7	5	7	7	7	7	7	7	7	7	7	3	7	7
C5	4	7	7	7	7	7	4	7	7	7	7	7	7	7	7	7	4	7	7
C6	5	5	7	7	7	5	6	7	7	7	7	6	7	7	7	7	5	7	7
C7	4	5	6	7	7	4	6	7	6	7	4	4	7	5	7	3	7	7	7
C1	7	4	5	6	5	5	5	3	6	4	5	4	5	5	4	4	5	4	5
C2	6	4	5	6	4	4	4	3	6	4	5	4	5	5	4	4	5	5	5
C3	6	3	5	4	4	4	4	3	4	4	4	4	3	4	3	5	5	5	4
C4	4	3	3	6	5	4	4	3	4	3	4	4	4	3	3	5	4	4	4
C5	4	4	3	6	5	4	3	3	4	3	4	4	4	2	4	5	3	3	4
C6	6	3	5	4	4	4	5	3	4	3	4	4	4	3	4	4	4	4	5
C7	6	4	5	4	4	3	5	3	3	2	4	4	3	4	4	3	4	5	5

TABLE 3.10
Data Matrix. Part 2

Criterio	Procesos																					
	P20	P21	P22	P23	P24	P25	P26	P27	P28	P29	P30	P31	P32	P33	P34	P35	P36	P37	P38	P39	P40	P41	P42
C1	3	6	4	7	3	6	4	7	7	7	7	6	4	5	2	4	4	4	4	4	6	6	7
C2	6	6	5	6	6	6	5	7	7	6	7	5	7	6	6	6	6	6	6	6	7	7	7
C3	4	4	7	5	2	4	6	4	5	6	5	4	3	6	7	7	7	7	7	7	4	4	7
C4	4	6	5	3	2	7	7	4	7	6	7	6	3	4	4	4	4	4	4	6	5	5	7
C5	3	5	5	4	4	7	7	4	6	6	7	4	3	5	5	5	5	5	5	7	4	4	7
C6	6	5	4	7	4	4	7	6	7	7	7	4	7	6	4	5	5	5	5	5	7	7	7
C7	6	5	4	7	4	4	7	6	7	7	7	4	7	7	4	4	4	4	4	4	7	7	7
C1	6	7	7	7	5	7	7	7	6	6	6	7	7	5	7	6	6	6	6	6	6	5	7
C2	5	5	5	5	3	5	5	5	6	5	5	6	6	5	5	5	5	5	5	5	5	5	5
C3	5	6	6	6	5	6	5	5	5	5	5	6	5	4	5	5	5	5	5	5	5	5	5
C4	5	5	5	6	4	5	5	5	6	5	5	6	6	5	5	5	5	4	5	5	5	5	5
C5	5	5	5	6	5	5	5	5	6	5	5	5	5	5	5	4	4	4	4	5	5	5	5
C6	6	5	5	5	4	5	6	5	5	5	5	5	5	5	5	5	5	5	5	5	5	5	5
C7	6	5	5	5	5	5	6	5	5	5	5	6	6	5	5	5	5	5	5	5	5	5	5
C1	6	7	6	4	2	7	3	7	6	6	7	5	7	3	4	3	3	3	6	3	5	5	5
C2	7	7	7	6	2	7	4	7	6	7	7	5	7	6	4	3	4	4	6	3	6	6	7
C3	7	7	7	7	3	6	7	4	7	7	7	7	5	6	7	7	7	7	7	3	6	6	7
C4	7	7	7	7	7	7	7	4	5	6	7	4	5	6	7	7	7	7	4	7	4	4	7
C5	6	7	7	7	7	6	7	4	5	7	7	4	5	6	7	7	7	7	6	7	4	4	7
C6	5	7	7	7	4	7	7	7	7	7	7	5	7	7	7	5	6	7	7	7	7	7	7
C7	7	7	4	4	3	3	6	4	7	7	7	7	5	7	6	3	5	3	6	3	7	7	7
C1	6	6	3	5	3	5	4	5	5	5	5	4	4	3	4	3	3	3	3	4	4	4	4
C2	4	6	4	5	3	5	4	4	4	4	3	3	5	4	4	4	4	4	4	3	3	3	4
C3	6	4	5	4	4	5	5	4	4	4	4	4	4	5	6	5	5	6	5	6	4	4	5
C4	4	4	5	4	4	5	4	3	3	5	4	3	4	4	4	4	4	3	4	4	3	3	3
C5	3	3	4	4	4	4	5	3	3	4	3	3	3	4	4	3	4	4	4	4	3	3	3
C6	6	4	6	5	4	4	4	3	3	4	4	3	3	4	4	6	4	3	4	4	3	3	4
C7	5	4	4	4	5	3	3	4	3	4	3	4	3	4	4	3	3	3	4	3	4	4	4

TABLE 3.10
Data Matrix. Part 3

	Procesos									Mecanismos Relacionales																
Criterio	**P43**	**P44**	**P45**	**P46**	**P47**	**P48**	**P49**	**P50**	**P51**	**M52**	**M53**	**M54**	**M55**	**M56**	**M57**	**M58**	**M59**	**M60**	**M61**	**M62**	**M63**	**M64**	**M65**	**M66**	**M67**	**M68**
C1	7	7	6	7	3	4	7	7	7	7	7	7	7	7	7	6	7	7	5	7	4	7	4	7	6	6
C2	6	6	4	6	6	5	7	7	7	6	7	7	7	7	7	6	7	7	5	6	4	6	5	7	6	6
C3	6	6	7	5	7	6	7	4	4	4	7	5	5	5	5	4	6	6	4	5	6	6	7	4	6	6
C4	4	4	4	4	4	4	7	7	7	7	7	7	7	7	5	7	7	7	6	7	7	6	5	6	6	6
C5	4	4	5	6	6	5	7	4	4	4	7	7	4	7	6	6	5	5	6	7	7	6	6	7	7	5
C6	6	6	5	5	6	5	7	7	7	7	7	7	7	7	7	6	7	7	6	7	4	5	6	6	4	5
C7	6	6	5	5	5	6	7	7	7	7	7	7	7	7	7	6	7	7	6	7	4	5	4	6	4	5
C1	7	7	7	7	6	6	5	5	5	6	7	6	6	5	6	6	5	5	7	7	5	5	6	6	6	5
C2	5	5	6	6	5	5	5	5	5	5	5	5	5	5	5	5	5	5	5	5	5	5	5	5	5	5
C3	5	6	6	6	5	5	5	5	5	5	5	5	5	5	5	5	5	5	5	5	5	5	5	6	5	5
C4	5	5	6	6	5	5	5	5	5	5	5	5	5	5	5	5	5	5	5	5	5	5	5	6	5	5
C5	5	5	6	5	5	5	5	5	5	5	5	5	5	5	5	5	5	5	5	5	5	5	5	6	5	5
C6	5	6	6	5	5	5	5	5	5	5	5	5	5	5	5	5	5	5	5	5	5	5	5	6	5	5
C7	6	6	6	5	5	5	5	5	5	5	5	5	5	5	5	5	5	5	5	5	5	5	5	6	5	5
C1	5	6	7	3	4	4	3	3	6	7	7	7	7	7	7	7	7	4	3	3	2	5	3	7	7	7
C2	7	7	6	5	4	5	7	7	7	7	7	7	7	7	7	7	7	7	3	3	2	5	6	7	7	7
C3	7	7	7	7	7	7	7	3	4	5	6	6	6	6	3	4	6	6	7	7	5	6	7	7	7	7
C4	6	6	7	7	7	7	7	7	7	7	7	7	7	6	6	6	4	7	7	7	7	6	7	7	7	7
C5	6	5	7	7	7	6	7	4	4	6	6	7	7	6	6	6	4	7	7	7	7	6	7	7	7	7
C6	7	7	7	7	6	7	7	7	7	7	7	7	7	7	6	7	7	7	7	7	6	6	7	7	7	7
C7	7	7	7	4	4	4	3	7	7	7	7	7	3	7	7	7	6	4	3	3	3	3	4	6	6	4
C1	4	4	3	5	4	4	3	5	6	5	5	5	6	4	6	6	5	6	4	5	4	4	3	4	5	4
C2	5	4	4	4	4	5	4	5	5	4	5	5	4	4	4	4	5	5	4	4	4	3	4	4	5	4
C3	6	3	6	4	6	5	5	4	4	3	4	4	5	4	5	4	5	4	4	4	4	4	4	4	4	5
C4	3	3	3	3	3	3	3	3	4	5	4	5	4	4	5	5	4	4	4	4	5	4	5	4	4	4
C5	3	3	4	4	3	3	4	3	4	4	4	3	5	4	5	4	4	5	5	5	5	5	5	4	4	4
C6	4	4	4	4	3	4	3	4	5	4	5	4	4	4	4	4	4	4	4	4	3	4	4	3	4	4
C7	4	5	3	4	3	4	3	4	3	4	5	4	4	4	5	4	4	4	4	4	3	3	3	4	3	3

Step 2: Verification of preconditions.

The necessary preconditions were verified before developing the PCA. For reliability analysis, *internal consistency* was measured using *Cronbach's Alpha* ($\alpha = 0.980$) and *McDonald's Omega* ($\omega = 0.981$) coefficients where both indicated an excellent level of reliability (George & Mallery, 2003).

Regarding the adequacy of the data, the *Kaiser-Meyer-Olkin (KMO)* statistical tests were applied which measures the degree of partial correlation between variables, indicating whether the values are appropriate for factor analysis (Phakiti, 2015; Loewen & Gonulal, 2015; Phakiti, 2015) and Bartlett's test of sphericity which contrasts the null hypothesis that the correlation matrix is an identity matrix (Dziuban & Shirkey, 1974). KMO showed an acceptable measure of adequacy (0.50) and Bartlett's test of sphericity showed a significant value ($p < 0.001$) indicating that the correlations are strong enough to apply PCA.

Step 3: Component extraction and determination of the number of factors.

The *PCA method* was applied to obtain the factors that concentrate the highest possible variance of the original variables. To represent how much of the total variance of the data set a component explains, an *eigenvalue* ≥2.0 was determined, which represents the variable explained by each component. To determine the number of factors to be retained, criteria such as Kaiser's rule, scree plot, and percentage of variance explained of the total variance were used.

Step 4: Factor rotation.

The *Varimax* rotation method was used to obtain a clearer and simpler structure in the factorial loadings to improve the interpretability of the components obtained, facilitating the coherent grouping of the variables around each component. The type of varimax rotation is appropriate for this study because it allows us to identify the key components of the GTI while preserving the independence between components, which is ideal for categorizing the governance dimensions without ambiguity.

Table 3.11 presents the values resulting from running the factor analysis with eigenvalue ≥2.0, with the varimax rotation method, a factor loading ≥0.5, resulting in a total of six principal components whose corresponding items can be observed in the same table, with their corresponding factor loadings. Likewise, Table 3.12 presents the calculation of the variance for each component.

TABLE 3.11
Principal Component Analysis with *Eigenvalue* ≥ 2.0

Items	1	2	3	4	5	6
E05	0.905					
M58	0.899					
M54	0.885					
M52	0.881					
M56	0.877					
E04	0.877					
P30	0.827					
P51	0.802					
E15	0.786					
P29	0.781					
E13	0.775					
M53	0.761					
M57	0.739					
P50	0.717					
P28	0.712		0.514			
E03	0.686					
M66	0.672					
P42	0.656					
E08	0.654					
M59	0.631		0.534			
M60	0.617	0.505				
E19	0.602					
P21	0.578					
E18	0.555		0.522			
M55	0.554			0.522		
E10	0.540					
P37		0.863				
M61		0.842				
P36		0.832				
M63		0.825				
P35		0.822				
P47		0.808				
M65		0.807				
M62		0.797				
P26		0.795				
P39		0.788				
P48		0.719				
P49		0.714				
P46		0.707				
P34		0.648				
M64		0.620				
E11		0.554				
P38			0.745			
P43			0.738			
P44			0.698			
P40	0.604		0.671			

(Continued)

TABLE 3.11
Principal Component Analysis with *Eigenvalue* ≥ 2.0

Ítems	1	2	3	4	5	6
P41	0.622		0.662			
E17			0.661			
P31			0.587			
P33		0.508	0.566			
P45			0.557			
E01				0.702		
P27				0.670		
E14				0.658		
E09				0.584		
E12				0.562		
P23		0.510		0.561		
E07				0.550		
P32				0.543		
P25					0.753	
M67					0.706	
P22					0.672	
M68					0.648	
E06		0.512			0.614	
E16		0.585			0.607	
P20						0.656
E02						0.652
P24		0.504				0.538

TABLE 3.12
Percentage of Variance by Component

Component	SC Loadings	% of Variance	% Accumulated
1	18.40	27.06	27.1
2	13.66	20.09	47.1
3	8.73	12.84	60.0
4	5.67	8.34	68.3
5	5.34	7.85	76.2
6	3.56	5.23	81.4

Step 5: Interpretation of the loadings matrix.

The rotated component matrix shows the correlation coefficients between each original item and each principal component. Each value in this matrix indicates how much an item contributes to the corresponding component. Factor loadings with a value ≥0.5 were considered significant and were analyzed to assign conceptual significance to each component. Items that had a strong influence on each component were identified to perform Step 6.

Step 6: Component labeling.

Finally, a conceptual label was assigned to each retained component based on the analysis of its component items and the values of their factor loadings. These labels represent the key areas identified for the ITG. For example, if several items related to risk management processes, training and awareness, compliance with regulations are grouped in the same component, it can be interpreted as *"Risk management and compliance"*. Table 3.13 shows the list with the six components identified and labeled.

TABLE 3.13
Key Areas of IT Governance from the Principal Component Perspective

Component	Label	Areas That Comprise It
1	Strategic leadership and collaborative governance in IT	Institutional leadership and strategic direction. Governance, collaboration, and stakeholder participation. Performance management and value measurement. Professional competencies and technical capabilities.
2	Risk management, security, and compliance in ITG.=	Compliance. Risk-oriented organizational culture. Formal policies. Integrated IT risk management and information security culture.
3	Alignment, strategic and organizational IT management.=	Value-oriented use and planning of IT resources and organizational decisions. General governance policies. Clear organizational structure, decision making, decisions, and roles. Strategic alignments between business and IT.
4	Illegal strategic infrastructure for ITG	Structural characteristics of the organization. Management of IT-related strategic resources. Legal and intellectual protection on IT elements.
5	ITG policy, communication, talent	Organizational communication. Use of IT governance frameworks. Staff training and skills. Policies and regulations.
6	IT infrastructure management, resources, and evaluation	Technological resources infrastructure. Organizational and IT performance evaluation. IT investment and efficiency. Risk management and security control.

Fuzzy TOPSIS was used to identify the hierarchy of each of the critical factors for ITG implementation (alternatives A1-A68) and PCA was used to identify and name each of the components that empirically group the 68 items. Finally, the *Decision Model for IT Governance implementation* was designed *with an approach based on the Fuzzy TOPSIS method and Principal Component Analysis* that shows in Table 3.14 the six extracted components that contribute to the efficient implementation of ITG with the corresponding hierarchical items based on the closeness coefficient.

3.4 DISCUSSION

The proposed model contributes to the theoretical aspect in the context of ITG knowledge, as well as to the practical aspect, offering a methodology and an applicable model for organizational decision making, oriented to the fulfillment of strategic objectives. The model provides IT managers with a prioritized and structured guide on which key areas and variables should be managed with greater emphasis to ensure effective governance aligned with institutional goals.

Contrasting the proposed model against traditional frameworks such as COBIT, ISO/IEC 38500 (Alreemy et al., 2016) and the OECD approach, it highlights its ability to integrate, prioritize, and reduce the complexity of ITG through an empirical and systematic methodology. Unlike existing models, which offer general guidelines or extensive structuring, this proposal identifies 68 critical variables validated by experts, organized into six key areas adapted to the organizational context, which allows greater flexibility compared to the areas suggested by the OECD (Erasmus & Marnewick, 2021).

3.5 CONCLUSIONS

The results obtained from the study demonstrate that it is possible to build a hybrid model that integrates the Fuzzy TOPSIS fuzzy multicriteria method and PCA, based on a data collection instrument. This model allowed the identification and prioritization of a set of critical variables that have a direct impact on the effective implementation of ITG, thus strengthening organizational decision-making processes. The validity of the proposed model was satisfactorily supported by statistical indicators of reliability, internal consistency, and sample adequacy, confirming its methodological robustness and the relevance of the hybrid approach adopted. As a result, the model represents a strategic tool that contributes to the alignment between technology management and organizational objectives, reinforcing the ITG as a key component of organizational performance.

This research opens directions for imminent studies both in the methodological and implementation areas. The study leaves for future research the analysis of the generalization of the methodology used for the model in the development of decision-making models in large, as well as small- and medium-sized organizations, which have different capacities in technical, financial, and human resources. The analysis of the capacity of the theoretical model was argued from a theoretical point of view, so its implementation in case studies is suggested in order to analyze the generalization capacity of the model and to carry out the necessary tests to validate its efficiency.

TABLE 3.14
Decision Model for IT Governance Implementation with an Approach Based on the Fuzzy TOPSIS Method and Principal Component Analysis

ITG Mechanism	Proximity Coefficient CC	Code	Factorial Loads					
			Component 1	Component 2	Component 3	Component 4	Component 5	Component 6
Estructure	0.7662	E10	0.5400					
	0.7197	E17			0.6610			
	0.7123	E08	0.6540					
	0.7012	E19	0.6020					
	0.6599	E05	0.9050					
	0.6564	E16		0.5850			0.6070	
	0.6513	E03	0.6860					
	0.6496	E02						0.6520
	0.6198	E07				0.5500		
	0.6145	E14				0.6580		
	0.5989	E06		0.5120			0.6140	
	0.5976	E12				0.5620		
	0.5752	E09				0.5840		
	0.5601	E11		0.5540				
	0.5445	E13	0.7750					
	0.5289	E15	0.7860					
	0.5222	E01				0.7020		
	0.4671	E04	0.8770					
	0.2853	E18	0.5550		0.5220			
Process	0.6859	P21	0.5780					
	0.6515	P29	0.7810					
	0.6249	P25					0.7530	

0.6237	**P51**	0.8020					
0.6156	**P43**			0.7380			
0.6151	**P42**	0.6560					
0.6077	**P23**		0.5100		0.5610		
0.5927	**P30**	0.8270					
0.5893	**P45**			0.5570			
0.5708	**P28**	0.7120		0.5140			
0.5589	**P22**					0.6720	
0.5538	**P20**						0.6560
0.5477	**P48**		0.7190				
0.5437	**P26**		0.7950				
0.5390	**P44**			0.6980			
0.5249	**P38**			0.7450			
0.5221	**P27**				0.6700		
0.5121	**P49**		0.7140				
0.5045	**P46**		0.7070				
0.4918	**P34**		0.6480				
0.4846	**P33**		0.5080	0.5660			
0.4818	**P36**		0.8320				
0.4699	**P32**				0.5430		
0.4631	**P40**	0.6040		0.6710			
0.4627	**P50**	0.7170					
0.4584	**P35**		0.8220				
0.4569	**P41**	0.6220		0.6620			
0.4552	**P47**		0.8080				
0.4502	**P37**		0.8630				
0.4441	**P31**			0.5870			
0.4006	**P39**		0.7880				
0.1713	**P24**		0.5040				0.5380

(Continued)

TABLE 3.14 *(Continued)*
Decision Model for IT Governance Implementation with an Approach Based on the Fuzzy TOPSIS Method and Principal Component Analysis

ITG Mechanism	Proximity Coefficient CC	Code	Factorial Loads					
			Component 1	Component 2	Component 3	Component 4	Component 5	Component 6
Relacional	0.7199	**M53**	0.7610					
	0.6903	**M55**	0.5540			0.5220		
	0.6731	**M59**	0.6310		0.5340			
	0.6712	**M57**	0.7390					
	0.6702	**M54**	0.8850					
	0.6586	**M58**	0.8990					
	0.6258	**M67**					0.7060	
	0.6149	**M60**	0.6170	0.5050				
	0.5937	**M52**	0.8810					
	0.5897	**M56**	0.8770					
	0.5820	**M66**	0.6720					
	0.5739	**M68**					0.6480	
	0.5030	**M62**		0.7970				
	0.4983	**M64**		0.6200				
	0.4899	**M65**		0.8070				
	0.4560	**M61**		0.8420				
	0.3367	**M63**		0.8250				

REFERENCES

Abdollahbeigi, B., & Salehi, F. (2020). The critical factors of IT governance and its impact on organizational performance in Malaysian manufacturing industry. *Serbian Journal of Management, 15*(1), 81–99.

Abu Kwaik, N., Sweis, R., Allan, B., & Sweis, G. (2023). Factors affecting risk management in industrial companies in Jordan. *Administrative Sciences, 13*(5), 132.

Ahn, Y. S., Kang, J. H., Cho, H. R., & Kim, M. J. (2007). An analysis of influence factors for IT governance: Focusing on ITA/EA functions. *Journal of Information Technology Services, 6*(2), 63–80.

AlGhamdi, S., Win, K. T., & Vlahu-Gjorgievska, E. (2020). Information security governance challenges and critical success factors: Systematic review. *Computers & Security, 99*, 102030.

Ali, S., & Green, P. (2005, November). Determinants of effective information technology governance: A study of IT intensity. In *Proceedings of the International IT Governance Conference*. University of Auckland.

Ali, S., & Green, P. (2012). Effective information technology (IT) governance mechanisms: An IT outsourcing perspective. *Information Systems Frontiers, 14*(2), 179–193.

Alreemy, Z., Chang, V., Walters, R., & Wills, G. (2016). Critical success factors (CSFs) for information technology governance (ITG). *International Journal of Information Management, 36*(6), 907–916.

Alshahrani, S., Bah, S., & Murad, M. A. A. A. (2022). Analysis and ranking of IT risk factors using F-TOPSIS. *International Journal of Advanced Computer Science and Applications, 13*(7), 440–447.

Ayağ, Z. (2021). A comparison study of fuzzy-based multiple-criteria decision-making methods to evaluating green concept alternatives in a new product development environment. International Journal of Intelligent Computing and Cybernetics, 14(3), 412–438.

Azhar, N. A., Radzi, N. A. M., & Ahmad, W. S. H. M. W. (2021). Multi-criteria decision making: A systematic review. *Recent Advances in Electrical & Electronic Engineering, 14*(8), 779–801.

Bruma, L. M. (2020). An approach for information security risk assessment in cloud environments. *Informatica Economica, 24*(2), 29–40.

Castellanos, W. S. (2023). Impact of information technology (IT) governance on business–IT alignment. *TEC EMPRESARIAL, 18*(1), 40–52.

Chen, S.J. and Hwang, C.L. (1992). Fuzzy Multiple Attribute Decision Making Methods and Applications, Springer-Verlag.

da Silva Leite, L., Coelho, M. A. S., Simões, J. C. F., Mariano, A. M., Monteiro, S. B. S., & Souza, J. C. F. (2021, June). Factors affecting the successful implementation of IT governance: A study using structural equations applied to the banking industry. In *Proceedings of the 16th Iberian Conference on Information Systems and Technologies (CISTI)* (pp. 1–6). IEEE.

Debreceny, R. S. (2013). Research on IT governance, risk, and value: Challenges and opportunities. *Journal of Information Systems, 27*(1), 129–135.

De Haes, S., & Van Grembergen, W. (2005, January). IT governance structures, processes and relational mechanisms: Achieving IT/business alignment in a major Belgian financial group. In *Proceedings of the 38th Annual Hawaii International Conference on System Sciences* (p. 237b). IEEE.

De Haes, S., & Van Grembergen, W. (2009). An exploratory study into IT governance implementations and its impact on business/IT alignment. *Information Systems Management, 26*(2), 123–137.

Durán, M., & Ignacio, M. (2015). Critical factors in the adoption of information technologies in microenterprises of the mechanical workshops in the commune of Chillán [Master's thesis], Bío-Bío University.

Dziuban, C., & Shirkey, E. (1974). When is a correlation matrix appropriate for factor analysis? Some decision rules. *Psychological Bulletin, 81*(6), 358–361.

Erasmus, W., & Marnewick, C. (2021). An IT governance framework for IS portfolio management. *International Journal of Managing Projects in Business*, *14*(3), 721–742.
Fathi, M. R., Matin, H. Z., Zarchi, M. K., & Azizollahi, S. (2011). The application of fuzzy TOPSIS approach to personnel selection for Padir Company, Iran. *Journal of Management Research*, *3*(2), 1–14.
Fauzi, R., & Mulyana, R. (2020). Key factors in improving the maturity of information technology governance: A case study of state-owned enterprise in Indonesia. *International Journal of Innovation in Enterprise System*, *4*(2), 60–71.
Ferguson, C., Green, P., Vaswani, R., & Wu, G. (2013). Determinants of effective information technology governance. *International Journal of Auditing*, *17*(1), 75–99.
George, D., & Mallery, P. (2003). *SPSS for Windows step by step: A simple guide and reference* (11.0 update, 4th ed.). Allyn & Bacon.
Jolliffe, I. T., & Cadima, J. (2016). Principal component analysis: A review and recent developments. *Philosophical Transactions of the Royal Society A*, *374*(2065), 20150202.
Kaur, J., Mohamed, N., & Ahlan, A. R. (2012, September). Modeling the impact of information technology governance effectiveness using partial least squares. In *2012 International Conference on Statistics in Science, Business and Engineering (ICSSBE)* (pp. 1–5). IEEE.
Kumar, R., & Pamucar, D. (2025). A comprehensive and systematic review of multi-criteria decision-making (MCDM) methods to solve decision-making problems: Two decades from 2004 to 2024. *Spectrum of Decision Making and Applications*, *2*(1), 178–197. https://doi.org/10.31181/sdmap21202524
Kurti, I., Barolli, E., & Sevrani, K. (2013). Critical success factors for business–IT alignment: A review of current research. *Romanian Economic and Business Review*, *8*(3), 79–96.

APPENDIX A

Critical Factors of ITG after SLR

TABLE A1
Critical Factors of ITG: Structure Mechanism

IT Governance Mechanism	Code	Variables	References
Structure	E01	Business alignment	(Kurti et al., 2013; Tu & Yuan, 2014)
	E02	Organizational demographics	(Mohamed & Kaur a/p Gian Singh, 2012; Yudatama et al., 2019; Abdollahbeigi & Salehi, 2020; da Silva Leite et al., 2021; Abu Kwaik et al., 2023)
	E03	Consolidation of ITG structures	(Nfuka & Rusu, 2011; Kaur et al., 2012; Asi et al., 2017; Abdollahbeigi & Salehi, 2020; Sirait, 2021; Castellanos, 2023)
	E04	Consolidation and participation of committees	(De Haes & Van Grembergen, 2004; Ali & Green, 2005; De Haes & Van Grembergen, 2009; Ali & Green, 2012; Ferguson et al., 2013; Nfuka & Rusu, 2013; Wu et al., 2015; Levstek et al., 2018; AlGhamdi et al., 2020; Singh & Alhulail, 2023)

(Continued)

TABLE A1 *(Continued)*
Critical Factors of ITG: Structure Mechanism

IT Governance Mechanism	Code	Variables	References
	E05	Consolidation of committee structure	(Kaur et al., 2012; Castellanos, 2023)
	E06	Consolidation of communication structure	(Ali & Green, 2012; Kaur et al., 2012; Ferguson et al., 2013; Nfuka & Rusu, 2013; Alreemy et al., 2016; Yudatama et al., 2019; AlGhamdi et al., 2020; Fauzi & Mulyana, 2020; da Silva Leite et al., 2021)
	E07	Business strategies for strategic IT changes	(Castellanos, 2023; da Silva Leite et al., 2021; Yudatama et al., 2019)
	E08	Business strategies for aligning the business with IT	(Castellanos, 2023; da Silva Leite et al., 2021; Debreceny, 2013; Kurti et al., 2013; Yudatama et al., 2019)
	E09	Principles and policies related to decision-making	(Castellanos, 2023; da Silva Leite et al., 2021; Debreceny, 2013; Yudatama et al., 2019)
	E10	Authority position in decision making	(Mohamed & Kaur a/p Gian Singh, 2012; Abdollahbeigi & Salehi, 2020; Singh & Alhulail, 2023)
	E11	Business strategies for IT control and oversight	(Castellanos, 2023; da Silva Leite et al., 2021; Yudatama et al., 2019)
	E12	Organizational structure for employee management and understanding of roles and responsibilities	(Debreceny, 2013; Tu & Yuan, 2014; Abu Kwaik et al., 2023)
	E13	Senior management and board of directors	(Ali & Green, 2005; Nfuka & Rusu, 2011; Yildirim et al., 2011; Ali & Green, 2012; Ferguson et al., 2013; Kurti et al., 2013; Nfuka & Rusu, 2013; Tu & Yuan, 2014; Abdollahbeigi & Salehi, 2020; Sirait, 2021; Abu Kwaik et al., 2023)
	E14	Policies for acquisition and use of IT resources	(Nfuka & Rusu, 2011; Alreemy et al., 2016; AlGhamdi et al., 2020; da Silva Leite et al., 2021; Singh & Alhulail, 2023)
	E15	Impact of the board of directors' experience in information technology	(Singh & Alhulail, 2023)
	E16	Compliance with standards and policies	(Yildirim et al., 2011; Yudatama et al., 2019; AlGhamdi et al., 2020)
	E17	Sufficient resources for managing IT governance frameworks	(Setyadji et al., 2023)
	E18	Commitment to funding for IT	(Setyadji et al., 2023)
	E19	Support from senior management	(Yildirim et al., 2011; Tu & Yuan, 2014; Yudatama et al., 2019; da Silva Leite et al., 2021)

TABLE A2
Critical Factors of ITG: Process Mechanism

IT Governance Mechanism	Code	Variables	References
Process	P20	Resources for strategic IT assets	(Tu & Yuan, 2014; Yudatama et al., 2019)
	P21	Formulation of policies designed for IT governance	(Yudatama et al., 2019)
	P22	Policy related to organizational elements	(Yudatama et al., 2019)
	P23	Policies related to IT principles and responsibilities	(da Silva Leite et al., 2021; Yudatama et al., 2019)
	P24	Copyright in IT governance	(AlGhamdi et al., 2020)
	P25	IT governance policies relating to decisions and compliance	(Yudatama et al., 2019)
	P26	Policies relating to best practices in IT governance	(De Haes & Van Grembergen, 2004; Yudatama et al., 2019)
	P27	Business/IT alignment policies relating to strategic decisions	(Debreceny, 2013; Kurti et al., 2013; Yudatama et al., 2019)
	P28	Activities relating to IT strategies	(Ali & Green, 2005; Sirait, 2021)
	P29	Organizational support	(da Silva Leite et al., 2021; Tu & Yuan, 2014; Yudatama et al., 2019)
	P30	Supervisory board	(Yudatama et al., 2019)
	P31	Change management and implementation	(De Haes & Van Grembergen, 2004; Alreemy et al., 2016; Yudatama et al., 2019; Fauzi & Mulyana, 2020; Abu Kwaik et al., 2023)
	P32	Delivery of IT value	(AAsi et al., 2017; Sirait, 2021)
	P33	Performance measures and benchmarks for ongoing evaluation	(De Haes & Van Grembergen, 2004; Nfuka & Rusu, 2011; Tu & Yuan, 2014; AlGhamdi et al., 2020; Sirait, 2021)
	P34	Security policies	(Yildirim et al., 2011; Tu & Yuan, 2014; Yudatama et al., 2019; AlGhamdi et al., 2020)
	P35	Define roles and risk assumption	(AlGhamdi et al., 2020; Fauzi & Mulyana, 2020)
	P36	Management of risks related to IT methodology	(Tu & Yuan, 2014; Yudatama et al., 2019; AlGhamdi et al., 2020; Fauzi & Mulyana, 2020; da Silva Leite et al., 2021)
	P37	Management of risks related to control and supervision of IT resources	(Yildirim et al., 2011; Tu & Yuan, 2014; Yudatama et al., 2019; AlGhamdi et al., 2020)
	P38	Management of risks related to assessment and analysis of IT strengths and weaknesses	(Tu & Yuan, 2014; Yudatama et al., 2019; AlGhamdi et al., 2020; Fauzi & Mulyana, 2020)

(Continued)

TABLE A2 ***(Continued)***
Critical Factors of ITG: Process Mechanism

IT Governance Mechanism	Code	Variables	References
	P39	Government rules and regulations	(Yildirim et al., 2011; AlGhamdi et al., 2020; Abu Kwaik et al., 2023)
	P40	IT infrastructure	(Ahn et al., 2007; Nfuka & Rusu, 2013; AlGhamdi et al., 2020; Fauzi & Mulyana, 2020)
	P41	IT infrastructure	(Nfuka & Rusu, 2011; Fauzi & Mulyana, 2020; Abu Kwaik et al., 2023)
	P42	Competitive IT professionals	(Kaur et al., 2012)
	P43	Resource allocation	(Yildirim et al., 2011; Tu & Yuan, 2014)
	P44	IT investment and investment efficiency business processes	(Ahn et al., 2007; Wu et al., 2015)
	P45	Alignment between security controls and business	(AlGhamdi et al., 2020)
	P46	Policies and procedures	(AlGhamdi et al., 2020)
	P47	Organization security architecture	(AlGhamdi et al., 2020)
	P48	Business continuity and recovery plan	(AlGhamdi et al., 2020)
	P49	Third-party audit	(AlGhamdi et al., 2020)
	P50	IT value and performance measurement	(AlGhamdi et al., 2020; Ali & Green, 2005; De Haes & Van Grembergen, 2004; Wu et al., 2015)
	P51	Measurement and performance reports	(AlGhamdi et al., 2020)

TABLE A3
Critical Factors of ITG: Mechanism

IT Governance Mechanism	Código	Variables	References
Mechanism	M52	Business policies and guidelines	(Ali & Green, 2005; Tu & Yuan, 2014; Yudatama et al., 2019; da Silva Leite et al., 2021; Sirait, 2021)
	M53	IT management/senior management understand business objectives and IT contribution	(Nfuka & Rusu, 2011)
	M54	Communication and partnership between IT and the business	(Nfuka & Rusu, 2011; Kaur et al., 2012; Alreemy et al., 2016; Levstek et al., 2018)

(Continued)

TABLE A3
Critical Factors of ITG: Mechanism

IT Governance Mechanism	Código	Variables	References
	M55	Alignment and dissemination of IT strategies and corporate strategies	(Nfuka & Rusu, 2011; Kurti et al., 2013; Asi et al., 2017; Sirait, 2021)
	M56	Commitment to key stakeholders and IT/business communication.=	(Ali & Green, 2005; Nfuka & Rusu, 2011; Yudatama et al., 2019; Abdollahbeigi & Salehi, 2020; AlGhamdi et al., 2020; Sirait, 2021)
	M57	Stakeholder involvement	(Nfuka & Rusu, 2013; Alreemy et al., 2016; Yudatama et al., 2019; Fauzi & Mulyana, 2020; Singh & Alhulail, 2023)
	M58	Communication between the board of directors and executive managers	(Yudatama et al., 2019)
	M59	Trust and collaboration between the business and IT	(Abu Kwaik et al., 2023; Kurti et al., 2013)
	M60	Effective communications	(Abu Kwaik et al., 2023; AlGhamdi et al., 2020)
	M61	Board commitments related to standards and policies	(Yudatama et al., 2019)
	M62	CEO support for related standards, policies, and commitments	(Yudatama et al., 2019)
	M63	Promotion of ethics and a culture of compliance	(Ali & Green, 2005; Ali & Green, 2012; Ferguson et al., 2013; Wu et al., 2015; AlGhamdi et al., 2020)
	M64	Organizational culture	(AAli & Green, 2005; Mohamed & Kaur a/p Gian Singh, 2012; Ferguson et al., 2013; Alreemy et al., 2016; Asi et al., 2017; Yudatama et al., 2019; AlGhamdi et al., 2020; Abu Kwaik et al., 2023)
	M65	Information security culture	(Yildirim et al., 2011; Tu & Yuan, 2014; AlGhamdi et al., 2020; Setyadji et al., 2023)
	M66	IT competence of the board	(Yildirim et al., 2011; Kurti et al., 2013; Tu & Yuan, 2014; Yudatama et al., 2019; da Silva Leite et al., 2021; Singh & Alhulail, 2023)
	M67	Competence in the application and use of ITG frameworks	(da Silva Leite et al., 2021; Setyadji et al., 2023; Yudatama et al., 2019)
	M68	Effective awareness and training of human resources in ITG	(Weill & Ross, 2004; Ali & Green, 2005; Nfuka & Rusu, 2011; Yildirim et al., 2011; Kurti et al., 2013; Tu & Yuan, 2014; Yudatama et al., 2019; Fauzi & Mulyana, 2020; da Silva Leite et al., 2021; Sirait, 2021; Abu Kwaik et al., 2023; Singh & Alhulail, 2023)

APPENDIX: REFERENCES

Asi, P., Rusu, L., & Vieru, D. (2017). The role of culture in IT governance five focus areas: A literature review. *International Journal of IT/Business Alignment and Governance (IJITBAG), 8*(2), 42–61.

Levstek, A., Hovelja, T., & Pucihar, A. (2018). IT governance mechanisms and contingency factors: Towards an adaptive IT governance model. *Organizacija, 51*(4), 286–310.

Loewen, S., & Gonulal, T. (2015). Exploratory factor analysis and principal components analysis. In L. Plonsky (Ed.), *Advancing quantitative methods in second language research* (pp. 182–212). Routledge.

Marulanda-Echeverry, C. E., López-Trujillo, F., & Valencia-Duque, F. J. (2017). Governance and IT management in public entities. *AD-minister, 31*, 75–92.

Mohamed, N., & Kaur a/p Gian Singh, J. (2012). A conceptual framework for information technology governance effectiveness in private organizations. *Information Management & Computer Security, 20*(2), 88–106.

Mooi, E. A., Sarstedt, M., & Mooi-Reci, I. (2018). *Principal component and factor analysis.* Springer.

Nfuka, E. N., & Rusu, L. (2011). The effect of critical success factors on IT governance performance. *Industrial Management & Data Systems, 111*(9), 1418–1448.

Nfuka, E. N., & Rusu, L. (2013). Critical success framework for implementing effective IT governance in Tanzanian public sector organizations. *Journal of Global Information Technology Management, 16*(3), 53–77.

OECD. (2021). *The e-leaders handbook on the governance of digital government.* OECD Publishing.

Page, M. J., Moher, D., Bossuyt, P. M., Boutron, I., Hoffmann, T. C., Mulrow, C. D., Shamseer, L., Tetzlaff, J. M., Akl, E. A., Brennan, S. E., Chou, R., & Glanville, J. (2021). PRISMA 2020 explanation and elaboration: Updated guidance and exemplars for reporting systematic reviews. *BMJ, 372*, n160. https://doi.org/10.1136/bmj.n160

Parroquín-Amaya, P., & Pérez-Olguín, I. (2025). Multicriteria model for the identification of critical factors in IT governance [doctoral dissertation], Universidad Autónoma of Ciudad Juárez.

Phakiti, A. (2015). Quantitative research and analysis. In B. Paltridge, A. Phakiti (Eds.), *Research methods in applied linguistics: A practical resource* (pp. 27–48). Bloomsbury Publishing.

Ramírez-Ochoa, D.-D., Pérez-Domínguez, L. A., Martínez-Gómez, E.-A., & Luviano-Cruz, D. (2022). PSO, a swarm intelligence-based evolutionary algorithm as a decision-making strategy: A review. *Symmetry, 14*(3), Article 455. https://doi.org/10.3390/sym14030455

Rethlefsen, M. L., Kirtley, S., Waffenschmidt, S., Ayala, A. P., Moher, D., Page, M. J., Koffel, J. B., & PRISMA-S Group. (2021). PRISMA-S: Preferred reporting items for systematic reviews and meta-analyses literature search extension. *Systematic Reviews, 10*, Article 39. https://doi.org/10.1186/s13643-020-01542-z

Roshandel, M., Miri-Nargesi, M., & Hatami-Shirkouhi, M. (2021). The TOPSIS of different ideal solution and distance formula of fuzzy soft set in multi criteria decision making. *International Journal of Academic Research in Economics and Management Sciences, 10*(2), 78–85.

Sahoo, S. K., & Goswami, S. S. (2023). A comprehensive review of multiple criteria decision-making (MCDM) methods: Advancements, applications, and future directions. *Decision Making Advances, 1*(1), 25–48.

Salih, M. M., Zaidan, B. B., Zaidan, A. A., & Ahmed, M. A. (2019). Survey on fuzzy TOPSIS state-of-the-art between 2007 and 2017. *Computers & Operations Research, 104*, 207–227. https://doi.org/10.1016/j.cor.2018.12.019

Schmitz, C., & Pape, S. (2020). LiSRA: Lightweight security risk assessment for decision support in information security. *Computers & Security*, 90, 101656.

Setyadji, A. E., Putrananda, A. R., Permadi, D. H., Nustara, R. I., Pratama, R. B., Masyhuda, T. A., & Hariyanti, E. (2023). Causes of ineffective implementation of IT governance in risk management: A systematic literature review. *JIKO (Jurnal Informatika dan Komputer)*, *6*(2), 88–96.

Singh, H. P., & Alhulail, H. N. (2023). Information technology governance and corporate boards' relationship with companies' performance and earnings management: A longitudinal approach. *Sustainability*, *15*(8), 6492.

Sirait, K. B. (2021). The interrelation of information technology governance and enterprise risk management to organizational performance: A review of empirical literature. In *RSF Conference Series: Business, Management and Social Sciences*, 1(5), 37. Research Synergy Foundation.

Tu, Z., & Yuan, Y. (2014, August). Critical success factors analysis on effective information security management: A literature review. In *Proceedings of the 20th Americas Conference on Information Systems (AMCIS)* (pp. 1–13). Association for Information Systems.

Van Grembergen, W., De Haes, S., & Guldentops, E. (2004). Structures, processes and relational mechanisms for IT governance. In W. Van Grembergen (Ed.), *Strategies for information technology governance* (pp. 1–36). IGI Global.

Vejseli, S., & Rossmann, A. (2017). The impact of IT governance on firm performance: A literature review. In *PACIS 2017 proceedings*.

Wang, T. (2022). A novel approach of integrating natural language processing techniques with fuzzy TOPSIS for product evaluation. *Symmetry*, *14*(1), 120.

Wang, T. C., & Chang, T. H. (2007). Application of TOPSIS in evaluating initial training aircraft under a fuzzy environment. *Expert Systems with Applications*, *33*(4), 870–880.

Weill, P., & Ross, J. W. (2004). *IT governance: How top performers manage IT decision rights for superior results*. Harvard Business Press.

Weill, P., & Ross, J. W. (2004). *IT governance: How top performers manage IT decision rights for superior results*. Harvard Business Press.

Wu, S. P. J., Straub, D. W., & Liang, T. P. (2015). How information technology governance mechanisms and strategic alignment influence organizational performance. *MIS Quarterly*, *39*(2), 497–518.

Yildirim, E. Y., Akalp, G., Aytac, S., & Bayram, N. (2011). Factors influencing information security management in small-and medium-sized enterprises: A case study from Turkey. *International Journal of Information Management*, *31*(4), 360–365.

Yudatama, U., Hidayanto, A. N., Nazief, B. A. A., & Phusavat, K. (2019). Data to model the effect of awareness on the success of IT governance implementation: A partial least squares structural equation modeling approach (PLS-SEM). *Data in Brief*, *25*, 104333. https://doi.org/10.1016/j.dib.2019.104333

4 Social Entrepreneurship Ranking of Mexican State Governments

An Analysis with the Ordered Weighted Average Operator

Brenda Dennís Valadez-Solana, Martín Isimayrt Huesca-Gastélum, Enrique Cruz-Domínguez, Ana Luz Ramos-Soto, and Ernesto León-Castro

4.1 INTRODUCTION

The growing economic competition, coupled with globalization, has led societies worldwide to redesign their strategies to adapt to a more challenging environment (Carraher et al., 2016). In this sense, social entrepreneurship (SE) represents a field that creates innovative solutions to social problems and contributes to social transformation (Adro & Fernandes, 2021). Furthermore, the importance of SE in developing countries lies in the emphasis on market activities and innovation that promote poverty reduction, well-being, and sustainable development (Berrones-Flemmig & Mauroner, 2019; Comini et al., 2012; Gonçalves et al., 2023; Popkova & Sergi, 2021). In addition, the SE has received increasing academic and governmental interest due to its importance in promoting the economic and social development of the territories (Christie & Honig, 2006; Félix, 2011) and for having the potential to address some of the more complex problems of societies, which arise from market and government insufficiencies or failures (Griffiths et al., 2013).

This is why SE has also been called solidarity capitalism: its scope covers both for-profit and non-profit companies, and its purpose is to achieve a social mission by implementing innovation processes and balancing market competitiveness (Carraher et al., 2016; De la Garza-Carranza et al., 2020; Scillitoe et al., 2018). Therefore, it is important to develop constructs to quantify SE and analyze its contribution (Turner et al., 2014), particularly in emerging countries where government support for SE is limited

DOI: 10.1201/9781003724902-5

or nonexistent (Berrones-Flemmig & Mauroner, 2019). However, very few empirical studies have measured and evaluated the incidence of SE (Lepoutre et al., 2011) due to the lack of consensus in the scientific literature and the complexity of this variable.

In this sense, this study aims to classify the SE of Mexican state governments using the ordered weighted average (OWA) operator and its extensions. This proposal presents a more flexible and adaptable way to measure this variable using these aggregation operators. To this end, 13 indicators were established based on the studies by Gonçalves et al. (2023), Griffiths et al. (2013), Hoogendoorn (2016), and Popkova and Sergi (2021). These researchers consider macro-level indicators that allow SE to be measured in various social contexts.

This chapter is structured as follows: Section 4.2 presents a review of the literature on SE. Section 4.3 details the measurement of SE among Mexico's state governments using the OWA operator. Section 4.4 presents the study's results. Finally, Section 4.5 summarizes the main conclusions and future lines of research.

4.2 A REVIEW OF THE LITERATURE ON THE MEASUREMENT OF SOCIAL ENTREPRENEURSHIP

Entrepreneurship has been studied since the mid-20th century and has become a relevant topic across disciplines such as economics, psychology, sociology, political science, history, and anthropology (Berrones-Flemmig & Mauroner, 2019; Terán-Yépez & Guerrero-Mora, 2020). The term SE first appeared in a scientific journal in 1964 (Rey-Martí et al., 2016). The fundamental purpose of SE is to create social value for the public good (Adro & Fernandes, 2021).

Although the fundamental objective is the creation of social value, a social enterprise also generates economic value in the form of income, ensuring its financial independence and sustainability (Dacin et al., 2011). Consequently, the importance of SE lies in its impact on economic and social development (Rey-Martí et al., 2016). Therefore, SE is an innovative activity that can occur across the non-profit, business, and government sectors; its primary purpose is to create social value (Austin et al., 2006; Dacin et al., 2011).

In this study, SE is understood as the process of creating social value expressed in the context of social or private organizations, carried out by one or several social entrepreneurs, which involves multiple actors whose purpose is the search for solutions to collective problems (Adro & Fernandes, 2021; Austin et al., 2006; Dacin et al., 2011; Dees & Anderson, 2006). That is, SE is viewed from a hybrid perspective, in which commercial and non-profit activities complement each other to address social needs and problems (Peris-Ortiz et al., 2016).

Also, SE represents a source of income generation, social innovation, and the creation of social value (Defourny & Nyssens, 2010) from an economic, social, and political perspective (Bacq & Janssen, 2011), hence its importance to governments everywhere. The SE has been studied from various approaches, at the micro and macro levels (Hoogendoorn, 2016). However, most of this literature focuses on micro-level analysis (Short et al., 2009) or qualitative studies (Kerlin, 2009).

Likewise, quantitative research on its measurement remains scarce (Capella-Peris et al., 2020; Villegas-Mateos & Amorós, 2019), and research on SE focused on social entrepreneurs and social companies, such is the case of Carraher et al. (2016), research where a scale of 11 items is established to measure the degree of SE of a social entrepreneur regarding the fulfillment of its social mission.

Some other studies have focused on SE in developed countries (Gupta et al., 2020), leaving a research gap on this phenomenon in developing nations (Gonçalves et al., 2023). Even so, few SE studies provide instruments or scales to measure the nature of SE (Capella-Peris et al., 2020). They can be helpful so that society, governments, and companies can replicate and manage them (Olinsson, 2017).

In this sense, one of the studies that presents a large-scale measurement of SE is that of Gonçalves et al. (2023), which proposes a model to measure SE in developing countries. The proposed model was conceived from the review bibliography of entrepreneurship models indexed in Web of Science and Scopus between 1945 and 2019.

This model includes four dimensions of SE measurement related to the individual and organizational levels: social entrepreneurial intention, social entrepreneurial orientation, processes, and results. In addition, Gonçalves et al. (2023) emphasize that income generation is an element of the results dimension that favors SE, as it enables financial sustainability and promotes social transformation.

Another relevant proposal is the work of Griffiths et al. (2013), who measure SE using findings from the Global Entrepreneurship Monitor (GEM) 2009 study, with a sample of 47 countries ranging from highly industrialized to very marginalized economies. As a result, the authors propose various indicators among which stand out: Gross Domestic Product (GDP), corruption incidence rate, participation of women and men in the labor force, education level, Gini coefficient, average tax rate, and level of well-being. These indicators were supported by contextual variables and macro-level influences in the SE's sociopolitical, economic, and cultural spheres.

The findings revealed that sociopolitical variables accounted for 76% of the variation in social entrepreneurial activity. The most significant single determinant of activity SE is the level of female workforce participation. As in the study by Griffiths et al. (2013), Hoogendoorn (2016) uses data from the GEM 2009 to explore the prevalence and determinants of SE at the macro level.

In the research, Hoogendoorn (2016) proposes GDP, subsidies, and the probability of crime and violence as indicators for measuring SE. The results of this study indicate that the three indicators of SE are positively and significantly associated with participation in social business activities. However, subsidies appear to have the greatest influence on SE participation in a country, as they favor the growth of emerging social companies.

In addition, Popkova and Sergi (2021) published an article that aimed to model financial risk management data in SE for sustainable development in emerging economies. The authors use a data source, "Social entrepreneurship in the global economy: from virtual scores to big data," to identify specific factors that trigger the development of SE potential in underdeveloped territories with emerging economies. Popkova and Sergi (2021) argue that expanding investments in social enterprises, increasing financial stimuli, and increasing investments in research and development (R&D) have a greater impact on SE in developing countries.

Regarding studies on SE using a multicriteria approach, Džunić et al. (2018) evaluated the contribution of SE to the employment of socially excluded groups in Serbia.

To do this, they surveyed 1,180 social companies that employed 10,189 workers. The methodological framework was based on descriptive statistics and a multicriteria decision-making model that combined two methods: entropy to enhance the weights of the criteria and TOPSIS to classify the alternatives and select the most suitable.

The selected criteria were: young workers, elderly workers, female workers, unskilled workers, workers with basic education, and workers who belong to vulnerable groups. For their part, the chosen alternatives represented different types of social enterprises: development agencies, foundations, business incubators, companies for people with disabilities, associations, and cooperatives. The results showed that the social enterprises that contribute most to the labor integration of socially disadvantaged groups are associations, cooperatives, and companies with disabled people, the latter being the most outstanding, as it employs 20% of the total workforce of social enterprises in Serbia.

For their part, Benjumea and Arango-Botero (2019) conducted a literature review on SE and, through an AHP multicriteria analysis, weighted criteria, and alternatives on a scale from 1 to 9 to evaluate the social and economic impact of SE. To do this, they selected the following alternatives: job creation, buildings, aid approach, and voluntary nature. Likewise, the chosen criteria were: social mission, sustainability, collective interest, generation of social value, learning, and performance. The results showed the importance of the voluntary nature and the SE's role in helping focus on the collective interest. Likewise, sustainability proved to be a relevant factor in creating jobs related to the SE.

Another investigation is that of Keleş Tayşir et al. (2023), which aimed to provide a comprehensive list of obstacles and solutions to the creation of social enterprises. To this end, the authors applied several multicriteria methods: Fuzzy Delphi (F-Delphi), Enhanced Fuzzy Weight Ratio Evaluation Analysis (IMF-SWARA), and Fuzzy Weighted Aggregate Sum Product Evaluation (F-WASPAS). The results showed 25 barriers and 12 solutions; the most relevant obstacles were the lack of financial resources, lack of experience in creating companies or social enterprises, insufficient income, unqualified human resources, and lack of work and management experience. For their part, the solutions that stood out were to increase the number of incubators and to prioritize social entrepreneurs.

In another order of ideas, in Mexico, the SE is a little-studied phenomenon (Félix, 2011). There is still little research that presents models to determine SE across different methodologies and sheds light on its development, measurement, and macro-level contribution. One of the studies found is that of De la Garza-Carranza et al. (2020), which identifies the personal characteristics and organizational competencies required of entrepreneurs in Mexican social companies and proposes recommendations that contribute to the sustainability of these companies and, consequently, to the social development of Mexico.

To this end, the authors developed a structural equation model and a measurement instrument, proposing creativity, the innovation process, product or service innovation, autonomy, and tasks and roles as variables. The results of this study emphasize the importance of social enterprises in developing strategies that benefit groups facing societal inequality. In addition, the research of Berrones-Flemmig and Marouner (2019) analyzed SE patterns in Mexican social companies. Through in-depth interviews with nine social entrepreneurs, the authors collected information on the type of entrepreneurship, motivation, frame of reference, vision for beneficiaries, and the role of SE associations. The research results showed that SE is motivated by the drive to make an area a better and safer place to live, reduce crime and violence, and, at the same time, educate people to participate in civil society. Likewise, alliances between individuals and communities, the public and private sectors, and local and global organizations are essential to facilitate the flow of information and participation and to reach integral solutions.

Another relevant study is that of Villegas-Mateos and Vázquez-Maguirre (2020), which analyzes SE ecosystems and their influence on the creation of social enterprises in five Mexican states: Guanajuato, Jalisco, Puebla, Querétaro, and San Luis Potosí. The authors applied the GEM National Expert Survey to 188 experts: 74 in highly marginalized regions and 114 in low-marginalization regions. The research results indicated that in regions of low marginalization, more favorable conditions are perceived for SE than in regions of high marginalization; thus, social entrepreneurs face greater challenges in creating social enterprises in peripheral regions.

Finally, these theoretical approaches are the most used to determine the degree or capacity of countries to contribute to the resolution of social problems that promote economic development. For this reason, we opted to use indicators extracted from these methodologies in this document. Likewise, a fuzzy logic technique will be used, specifically the OWA operator, as it is an ideal tool for classifying and ranking state governments by their capacity for SE.

4.3 THE ORDERED WEIGHTED AVERAGE OPERATOR

An operator for aggregating information is the OWA operator introduced by Yager (1988). This operator aggregates information between the maximum and the minimum, and since then, many applications have been made (Beliakov et al., 2007; Yager & Kacprzyk, 2012). The definition is as follows.

Definition 1. An OWA operator of dimension n is a mapping OWA: $R^n \rightarrow R$ with an associated weight vector W of dimension n such that $\sum_{j=1}^{n} w_j = 1$ and $w_j \in [0,1]$, according to the following formula:

$$\text{OWA}(a_1, a_2, \ldots, a_n) - \sum_{j=1}^{n} w_j b_j, \tag{4.1}$$

where b_j is the jth largest element of the collection a_i.

Note that we can distinguish between the descending OWA (DOWA) and the ascending OWA (AOWA) operator. This difference is related by $w_j = w^*_{n-j+1}$, where w_j is the jth weight of the DOWA operator and w^*_{n-j+1} the jth weight of the AOWA operator.

4.4 MEASUREMENT OF SOCIAL ENTREPRENEURSHIP OF STATE GOVERNMENTS OF MEXICO WITH THE OWA OPERATOR

To classify Mexican state governments using the OWA operator, follow these steps.

Step 1. Given that the objective is to classify federal government entities based on their capacity for SE, this will be the main element and will include 13 criteria defined in accordance with the theoretical framework presented in Section 4.2. In addition, a new column to indicate whether higher is better (H) or lower is better (L). This information is presented in Table 4.1.

TABLE 4.1
Indicators to Measure the Social Entrepreneurship of State Governments

ID	Explanation	Category	H or L
CIR	Crime incidences	Indicator	L
TSS	Total spending on subsidies (in millions)	Indicator	H
GDP	Gross domestic product (in millions)	Indicator	H
CIR	Corruption incidence rate	Indicator	L
PWL	Percentage of participation of women in the labor force	Indicator	H
PML	Percentage of participation of men in the labor force	Indicator	H
EHL	Enrollment in higher education	Indicator	H
GIC	Gini coefficient	Indicator	L
MHI	Median household income (in MXN)	Indicator	H
ATR	Tax payment (in millions)	Indicator	H
TSF	Total spending on financing (in millions)	Indicator	H
TSR	Total spending on research and development	Indicator	H

Step 2. For each indicator proposed in Table 4.1, information was obtained from each Mexican State (see Table 4.2). The data are presented in Table 4.3.

Because the information has different values, it will be normalized to a range of 10 (maximum) to 1 (minimum). To do this, the following formulation

a. If the indicator is H, the formula is: $9 \times \left(\dfrac{score - minimum}{maximum - minimum} \right) + 1$

b. If the indicator is L, the formula is: $-9 \times \left(\dfrac{score - minimum}{maximum - minimum} \right) + 10$

The normalized results are presented in Table 4.4.

TABLE 4.2
State Governments of Mexico

ID	State Government
A1	Aguascalientes
A2	Baja California
A3	Baja California Sur
A4	Campeche
A5	Chiapas
A6	Chihuahua
A7	Ciudad de México
A8	Coahuila
A9	Colima
A10	Durango
A11	Estado de México
A12	Guanajuato
A13	Guerrero
A14	Hidalgo
A15	Jalisco
A16	Michoacán
A17	Morelos
A18	Nayarit
A19	Nuevo León
A20	Oaxaca
A21	Puebla
A22	Querétaro
A23	Quintana Roo
A24	San Luis Potosí
A25	Sinaloa
A26	Sonora
A27	Tabasco
A28	Tamaulipas
A29	Tlaxcala
A30	Veracruz
A31	Yucatán
A32	Zacatecas

TABLE 4.3
Results for Each Identifier for Each State Government

ID	A1	A2	A3	A4	A5	A6	A7	A8	A9
CIR	29,584	31,690	21,576	18,610	16,386	26,432	45,336	24,418	26,565
TSS	16,735	41,220	11,842	8,754	58,185	43,171	-	20,202	12,163
GDP	206,715	599,221	141,176	462,353	268,174	566,203	2,944,607	567,405	101,361
CIR	19,821	30,567	8,226	20,586	16,698	15,368	25,771	27,589	8,920
PWL	40	41	38	37	43	42	41	36	43
PML	60	59	62	63	57	58	59	64	57

(Continued)

TABLE 4.3 ***(Continued)***
Results for Each Identifier for Each State Government

ID	A1	A2	A3	A4	A5	A6	A7	A8	A9
EHL	54,196	128,620	22,590	29,520	81,960	129,563	551,879	112,943	21,608
GIC	0.3951	0.4212	0.4314	0.4682	0.4694	0.4421	0.4468	0.3928	0.4015
MHI	78,288	88,912	91,417	57,458	39,845	81,929	89,310	75,126	69,321
ATR	1,599	6,764	1,333	1,681	1,840	6,689	-	3,450	901
TSF	544	12,526	600	210	765	4,150	-	3,662	1,240
TSR	475,889	1,787,047	444,711	323,277	705,629	1,025,624	14,757,497	956,702	372,507

ID	A10	A11	A12	A13	A14	A15	A16	A17	A18
CIR	17,387	45,501	29,936	26,481	19,205	31,944	18,102	32,059	21,214
TSS	14,593	160,921	37,526	43,321	34,099	46,452	25,680	22,694	16,675
GDP	200,397	1,575,492	682,552	227,408	251,438	1,186,919	406,943	186,362	116,192
CIR	50,885	32,906	24,185	24,373	11,849	16,497	29,716	23,345	19,651
PWL	39	42	41	48	43	42	45	46	46
PML	61	58	59	52	57	58	55	54	54
EHL	51,530	449,396	160,622	73,097	106,502	256,784	114,527	63,162	33,805
GIC	0.4521	0.4540	0.3856	0.4453	0.4013	0.3884	0.4045	0.4156	0.4076
MHI	57,220	57,233	60,100	41,754	53,234	71,744	56,959	57,239	65,326
ATR	2,237	22,786	6,092	1,225	1,638	6,425	2,150	770	931
TSF	3,707	24,652	5,720	2,100	1,168	5,757	4,128	6,412	29
TSR	398,763	2,988,258	1,811,662	316,713	863,165	3,257,382	1,416,182	1,857,610	283,893

ID	A19	A20	A21	A22	A23	A24	A25	A26	A27
CIR	34,099	18,552	36,234	31,817	31,538	29,122	30,230	30,200	35,448
TSS	56,487	52,830	33,548	29,434	21,258	31,038	40,202	43,330	21,418
GDP	1,344,947	249,591	542,719	388,481	257,699	354,896	383,066	572,000	502,306
CIR	18,684	27,484	37,896	21,849	31,389	18,732	21,065	31,883	25,277
PWL	36	50	43	41	41	41	40	40	37
PML	64	50	57	59	59	59	60	60	63
EHL	240,776	73,620	262,342	84,760	38,816	79,524	131,685	113,588	71,109
GIC	0.4550	0.4532	0.4386	0.4214	0.4623	0.4398	0.3955	0.4425	0.4557
MHI	86,018	43,343	49,384	74,956	71,902	60,144	71,888	75,270	52,095
ATR	11,879	1,509	5,284	3,407	4,151	2,065	2,805	4,326	2,262
TSF	10,275	2,853	-	650	4,298	909	154	5,672	5,269
TSR	2,514,009	598,964	2,095,555	1,447,360	259,278	1,260,287	971,471	1,322,645	474,248

ID	A28	A29	A30	A31	A32
CIR	20,473	28,718	19,545	23,600	22,474
TSS	22,115	14,906	46,244	14,643	12,151
GDP	481,559	94,889	767,013	259,829	150,782
CIR	14,571	30,007	39,296	19,495	12,219
PWL	40	46	42	42	39
PML	60	54	58	58	61
EHL	135,883	39,408	186,215	83,665	49,236
GIC	0.4260	0.3947	0.4196	0.4654	0.4494
MHI	63,521	46,295	44,637	62,371	50,136
ATR	5,026	623	5,286	2,499	2,359
TSF	4,576	-	3,935	360	148
TSR	639,989	290,457	1,509,718	1,257,005	500,504

TABLE 4.4
Normalized Results for the Different State Governments

ID	A1	A2	A3	A4	A5	A6	A7	A8	A9
CIR	5.92	5.27	8.40	9.31	10.00	6.89	1.05	7.52	6.85
TSS	1.94	3.31	1.66	1.49	4.25	3.41	1.00	2.13	1.68
GDP	1.35	2.59	1.15	2.16	1.55	2.49	10.00	2.49	1.02
CIR	7.55	5.29	10.00	7.39	8.21	8.49	6.30	5.91	9.85
PWL	3.38	4.28	2.48	1.96	5.50	4.66	4.21	1.26	5.56
PML	7.62	6.72	8.52	9.04	5.50	6.34	6.79	9.74	5.44
EHL	1.55	2.82	1.02	1.13	2.02	2.83	10.00	2.55	1.00
GIC	8.99	6.18	5.09	1.13	1.00	3.94	3.43	9.23	8.29
MHI	7.71	9.56	10.00	4.07	1.00	8.34	9.63	7.16	6.14
ATR	1.63	3.67	1.53	1.66	1.73	3.64	1.00	2.36	1.36
TSF	1.20	5.57	1.22	1.08	1.28	2.52	1.00	2.34	1.45
TSR	1.13	1.95	1.12	1.04	1.28	1.48	10.00	1.43	1.07

ID	A10	A11	A12	A13	A14	A15	A16	A17	A18
CIR	9.69	1.00	5.81	6.88	9.13	5.19	9.47	5.16	8.51
TSS	1.82	10.00	3.10	3.42	2.91	3.60	2.44	2.27	1.93
GDP	1.33	5.68	2.86	1.42	1.49	4.45	1.99	1.29	1.07
CIR	1.00	4.79	6.63	6.59	9.24	8.26	5.47	6.81	7.59
PWL	3.12	5.18	4.34	8.84	5.82	4.66	6.79	7.24	7.24
PML	7.88	5.82	6.66	2.16	5.18	6.34	4.21	3.76	3.76
EHL	1.51	8.26	3.36	1.87	2.44	4.99	2.58	1.71	1.21
GIC	2.86	2.65	10.00	3.59	8.32	9.70	7.97	6.78	7.64
MHI	4.03	4.03	4.53	1.33	3.34	6.57	3.99	4.04	5.45
ATR	1.88	10.00	3.41	1.48	1.65	3.54	1.85	1.30	1.37
TSF	2.35	10.00	3.09	1.77	1.43	3.10	2.51	3.34	1.01
TSR	1.09	2.69	1.96	1.04	1.37	2.86	1.72	1.99	1.02

ID	A19	A20	A21	A22	A23	A24	A25	A26	A27
CIR	4.52	9.33	3.86	5.23	5.32	6.06	5.72	5.73	4.11
TSS	4.16	3.95	2.88	2.65	2.19	2.74	3.25	3.42	2.20
GDP	4.95	1.49	2.41	1.93	1.51	1.82	1.91	2.51	2.29
CIR	7.79	5.94	3.74	7.13	5.11	7.78	7.29	5.01	6.40
PWL	1.00	10.00	5.82	4.34	4.09	4.54	3.51	3.70	1.51
PML	10.00	1.00	5.18	6.66	6.91	6.46	7.49	7.30	9.49
EHL	4.72	1.88	5.09	2.07	1.29	1.98	2.87	2.56	1.84
GIC	2.55	2.75	4.31	6.16	1.77	4.18	8.94	3.90	2.47
MHI	9.06	1.61	2.66	7.13	6.59	4.54	6.59	7.18	3.14
ATR	5.69	1.60	3.09	2.35	2.64	1.82	2.11	2.71	1.89
TSF	4.75	2.04	1.00	1.24	2.57	1.33	1.06	3.07	2.92
TSR	2.40	1.21	2.14	1.74	1.00	1.62	1.44	1.66	1.13

ID	A28	A29	A30	A31	A32
CIR	8.74	6.19	9.02	7.77	8.12
TSS	2.24	1.83	3.59	1.82	1.68
GDP	2.22	1.00	3.12	1.52	1.18
CIR	8.66	5.40	3.44	7.62	9.16
PWL	3.83	7.36	4.86	5.05	3.19
PML	7.17	3.64	6.14	5.95	7.81

(Continued)

TABLE 4.4 ***(Continued)***
Normalized Results for the Different State Governments

ID	A28	A29	A30	A31	A32
EHL	2.94	1.30	3.79	2.05	1.47
GIC	5.66	9.03	6.35	1.43	3.15
MHI	5.13	2.13	1.84	4.93	2.80
ATR	2.99	1.25	3.09	1.99	1.93
TSF	2.67	1.00	2.44	1.13	1.05
TSR	1.24	1.02	1.78	1.62	1.15

Step 3. The relative importance of each criterion was obtained by asking three different experts about the topic. The experts' profiles included 2 academics with more than 10 years of experience in SE, and another was a government worker with more than 15 years of experience in social programs and related areas. After all the weights were obtained, they were unified using the mean (see Table 4.5).

TABLE 4.5
Relative Weights for the Criterion

Criteria	Weight (%)
CIR	2.35
TSS	14.05
GDP	13.80
CIR	5.68
PWL	7.49
PML	8.23
EHL	6.41
GIC	3.88
MHI	12.64
ATR	5.13
TSF	15.73
WEL	4.61
TSR	2.35

Step 4. The average, weighted average (WA), OWA_{min}, OWA_{max} and interval numbers (IN) with a 4-tuplet using the formula $\frac{(c_1+2c_2+2c_3+c_4)}{6}$ (León-Castro et al., 2018; Moore, 1966) are calculated (see Table 4.6).

TABLE 4.6
Performance of State Governments Using Different Aggregation Operators

Operator	A1	A2	A3	A4	A5	A6	A7	A8	A9
Average	4.16	4.77	4.35	3.46	3.61	4.59	5.37	4.51	4.14
Weighted Average	3.65	4.90	3.89	2.97	3.02	4.38	5.44	4.03	3.47
OWA min	2.81	3.79	2.77	2.23	2.35	3.52	3.53	3.17	2.71
OWA max	5.68	5.76	6.20	4.96	5.07	5.76	7.23	6.05	5.67
Interval Number	4.02	4.82	4.24	3.34	3.45	4.54	5.40	4.38	3.94
Operator	A10	A11	A12	A13	A14	A15	A16	A17	A18
Average	3.21	5.84	4.65	3.37	4.36	5.27	4.25	3.81	3.98
Weighted Average	2.84	6.70	4.14	2.88	3.49	4.91	3.55	3.42	3.24
OWA min	2.17	4.40	3.69	2.31	3.01	4.34	3.13	2.79	2.63
OWA max	4.42	7.37	5.72	4.60	5.86	6.28	5.52	4.90	5.48
Interval Number	3.12	6.14	4.50	3.24	4.09	5.16	4.04	3.69	3.76
Operator	A19	A20	A21	A22	A23	A24	A25	A26	A27
Average	5.13	3.57	3.52	4.05	3.42	3.74	4.35	4.06	3.28
Weighted Average	5.41	3.04	3.15	3.71	3.38	3.33	3.85	4.05	3.20
OWA min	3.92	2.37	2.85	3.00	2.47	2.76	3.10	3.25	2.38
OWA max	6.41	5.02	4.20	5.17	4.45	4.79	5.68	4.97	4.32
Interval Number	5.24	3.43	3.40	3.95	3.42	3.62	4.20	4.07	3.28
Operator	A28	A29	A30	A31	A32				
Average	4.46	3.43	4.12	3.57	3.56				
Weighted Average	3.88	2.67	3.56	3.11	2.88				
OWA min	3.32	2.24	3.22	2.52	2.37				
OWA max	5.72	4.82	5.14	4.78	4.95				
Interval Number	4.29	3.21	3.95	3.44	3.36				

To better understand the results, a ranking based on the different operators was prepared (see Annex 4.1). The main idea of using different aggregation operators and reordering the arguments and weights is to visualize the ranking's sensitivity to changes in the relative importance of each indicator. From the results presented in Annex 1, it is evident that A11 is the best State across all operators, indicating that, even when different criteria are applied, Estado de Mexico remains the best in SE. After that, it can be seen that even the number 2 in the ranking can change: in 4 operators, it is A7, but in the OWAmin it drops to 6, and A15, which was 3, changes to 2. Another case is when the bottom score is analyzed, as the worst State is A10 (Durango) across 3 of the 5 operators. But the case of A27 (Tabasco) is very interesting because it ranks 32 on average, but can reach 23 in the WA rank and 25 in the OWA min, making a significant change in the understanding of SE in the State.

4.5 CONCLUSIONS

This chapter presents a classification of SE of the state governments of Mexico through the OWA operator. A fuzzy systems tool that allowed the different levels of SE to be compared based on different weighting vectors. This analysis demonstrated that, using the OWA operator, it is possible to obtain a ranking of Mexican states' governments based on the importance of each criterion.

The results indicate that A11 is the federal entity with the highest evaluation, while A10 is the state task with the lowest evaluation. In this sense, using this instrument represents an opportunity for policymakers because it allows them to identify the factors that require greater attention, recognize the areas with the best performance, and allocate resources to the areas of opportunity.

For future research, more complex aggregation operators based on the OWA operator should be applied, such as the heavy OWA (León-Castro et al., 2018) or the prioritized OWA (Pérez-Arellano et al., 2019). In addition, it is necessary to conduct a broader analysis using data from different municipal governments and other countries worldwide.

APPENDIX

ANNEX 4.1
Ranking Using Different Aggregation Operators

Operator	Average Rank	WA Rank	OWA Min Rank	OWA Max Rank	IN Rank
A11	1.00	1.00	1.00	1.00	1.00
A7	2.00	2.00	6.00	2.00	2.00
A15	3.00	4.00	2.00	4.00	4.00
A19	4.00	3.00	3.00	3.00	3.00
A2	5.00	5.00	4.00	8.00	5.00
A12	6.00	7.00	5.00	10.00	7.00
A6	7.00	6.00	7.00	9.00	6.00
A8	8.00	9.00	11.00	6.00	8.00
A28	9.00	11.00	8.00	11.00	9.00
A14	10.00	17.00	14.00	7.00	12.00
A25	11.00	12.00	13.00	12.00	11.00
A3	12.00	10.00	19.00	5.00	10.00
A16	13.00	16.00	12.00	15.00	14.00
A1	14.00	14.00	17.00	13.00	15.00
A9	15.00	18.00	21.00	14.00	18.00
A30	16.00	15.00	10.00	18.00	16.00
A26	17.00	8.00	9.00	21.00	13.00
A22	18.00	13.00	15.00	17.00	17.00
A18	19.00	22.00	22.00	16.00	19.00
A17	20.00	19.00	18.00	24.00	20.00
A24	21.00	21.00	20.00	26.00	21.00
A5	22.00	27.00	28.00	19.00	22.00

(Continued)

ANNEX 4.1 ***(Continued)***
Ranking Using Different Aggregation Operators

Operator	Average Rank	WA Rank	OWA Min Rank	OWA Max Rank	IN Rank
A31	23.00	25.00	23.00	27.00	23.00
A20	24.00	26.00	27.00	20.00	24.00
A32	25.00	30.00	26.00	23.00	27.00
A21	26.00	24.00	16.00	32.00	26.00
A4	27.00	28.00	31.00	22.00	28.00
A29	28.00	32.00	30.00	25.00	31.00
A23	29.00	20.00	24.00	29.00	25.00
A13	30.00	29.00	29.00	28.00	30.00
A27	31.00	23.00	25.00	31.00	29.00
A10	32.00	31.00	32.00	30.00	32.00

REFERENCES

Adro, F.D., & Fernandes, C. (2021). Social entrepreneurship and social innovation: Looking inside the box and moving out of it. Innovation: The European Journal of Social Science Research, 1–27.

Austin, J., Stevenson, H., & Wei-Skillern, J. (2006). Social and commercial entrepreneurship: same, different, or both? Entrepreneurship Theory and Practice, 30(1), 1–22.

Bacq, S., & Janssen, F. (2011). The multiple faces of social entrepreneurship: A review of definitional issues based on geographical and thematic criteria. Entrepreneurship & Regional Development, 23(5–6), 373–403.

Beliakov, G., Pradera, A., & Calvo, T. (2007). Aggregation functions a guide for practitioners (vol. 221). SPRINGER, Switzerland.

Benjumea, M.L., & Arango-Botero, D. (2019). Evaluating the impact of social entrepreneurship: A multicriteria approach. Periodica Polytechnica Social and Management Sciences, 27(2), 141–153.

Berrones-Flemmig, C., & Mauroner, O. (2019). Understanding social entrepreneurship in developing countries - a multiple case study from Mexico. International Journal of Business and Globalisation, 22(3), 389.

Capella-Peris, C., Gil-Gómez, J., Martí-Puig, M., & Ruíz- Bernardo, P. (2020). Development and validation of a scale to assess social entrepreneurship competency in higher education development and validation of a scale to assess social. Journal of Social Entrepreneurship, 11(1), 1–17.

Carraher, S.M., Welsh, D.H.B., Svilokos, A. (2016). Validation of a measure of social entrepreneurship. European Journal of International Management, Special Issue on Internationalization of Social Entrepreneurship, 10(4), 386–402.

Christie, M.J., & Honig, B. (2006). Social entrepreneurship: New research findings. Journal of World Business, 41(1), 1–5.

Comini, G., Barki, E., & Aguiar, L.T. de. (2012). A three-pronged approach to social business: A Brazilian multi-case analysis social businesses. Revista de Administração, 47(3), 385–397.

Dacin, T., Dacin, P., & Tracey, P. (2011). Social entrepreneurship: A critique and future directions. Organization Science, 22(5), 1203–1213.

Dees, G., & Anderson, B.B. (2006). Framing a theory of social entrepreneurship: Building on two schools of practice and thought. En R. Mosher-Williams (Ed.), Research on social entrepreneurship: Understanding and contributing to an emerging field. ARNOVA Occasional Paper Series, 1(3), (pp. 39–66).

Defourny, J., & Nyssens, M. (2010). Conceptions of social enterprise and social entrepreneurship in Europe and the United States: Convergences and divergences. Journal of Social Entrepreneurship, 1(1), 32–53.

De la Garza-Carranza, M.T., Guzmán-Soria, E., López Lemus, J.A., & Sierra-Martínez, A.C. (2020). Social entrepreneurship innovation: A study from Mexico. Entrepreneurial Business and Economics Review, 8(4), 201–220.

Džunić, M., Stanković, J., & Janković-Milić, V. (2018). Multicriteria approach in evaluating the contribution of social entrepreneurship to the employment of socially excluded groups. Technological and Economic Development of Economy, 24(5), 1885–1908.

Félix, M. (2011). Social entrepreneurship in Mexico: A conceptual analysis. International Journal of Business Competition and Growth, 1(4), 314.

Gonçalves, V.V., Oliveira, V.M., & Miki, A.F.C. (2023). Social entrepreneurship measurement framework for developing countries. *Revista de Administração Contemporânea, 27*(2), e220017.

Griffiths, M.D., Gundry, L.K., & Kickul, J.R. (2013). The socio-political, economic, and cultural determinants of social entrepreneurship activity. Journal of Small Business and Enterprise Development, 20(2), 341–357.

Gupta, P., Chauhan, S., Paul, J., & Jaiswal, M.P. (2020). Social entrepreneurship research: A review and future research agenda. Journal of Business Research, 113, 209–229.

Hoogendoorn, B. (2016). The prevalence and determinants of social entrepreneurship at the macro level. Journal of Small Business Management, 54, 278–296.

Keleş Tayşir, N., Ülgen, B., & İyigün, N.Ö. (2023). A framework to overcome barriers to social entrepreneurship using a combined fuzzy MCDM approach. Soft Computing, 28, 2325–2351.

Kerlin, J.A. (2009). Social Enterprise: A Global Comparison. Tufts University Press, Medford.

León-Castro, E., Avilés-Ochoa, E., Merigó, J., & Gil-Lafuente, A. (2018). Heavy moving average and their application in econometric forecasting. Cybernetics and Systems, 49(1), 26–43.

Lepoutre, J., Justo, R., Terjesen, S., & Bosma, N. (2011). Designing a global standardized methodology for measuring social entrepreneurship activity: The global entrepreneurship monitor social entrepreneurship study. Small Business Economics, 40(3), 693–714.

Moore, R. E. (1966). Interval Analysis. Prentice-Hall

Olinsson, S.B. (2017). Social entrepreneurship-committing theory to practice. Journal of Social Entrepreneurship, 8(2), 225–247.

Pérez-Arellano, L., León-Castro, E., Avilés-Ochoa, E., & Merigó, J. (2019). Prioritized induced probabilistic operator and its application in group decision making. International Journal of Machine Learning and Cybernetics, 10(3), 451–462.

Peris-Ortiz, M., Rueda-Armengot, C., & Palacios-Marqués, D. (2016). Is it possible to measure social entrepreneurship in firms? Cuadernos de Gestion, 16(2), 15–28.

Popkova, E.G., & Sergi, B.S. (2021). Dataset modelling of the financial risk management of social entrepreneurship in emerging economies. Risks, 9, 211.

Rey-Martí, A., Ribeiro-Soriano, D., & Palacios-Marqués, D. (2016). A bibliometric analysis of social entrepreneurship. Journal of Business Research, 69, 1651–1655.

Scillitoe, J.L., Poonamallee, L., & Joy, S. (2018). Balancing market versus social strategic orientations in socio-tech ventures as part of the technology innovation adoption process-examples from the global healthcare sector. Journal of Social Entrepreneurship, 9(3), 257–287.

Short, J.C., Moss, T.W., & Lumpkin, G.T. (2009). Research in social entrepreneurship: Past contributions and future opportunities. Strategic Entrepreneurship Journal, 3(2), 161–194.

Terán-Yépez, E.F., & Guerrero-Mora, A.M. (2020). Teorías de emprendimiento: revisión crítica de la literatura y sugerencias para futuras investigaciones. Revista Espacios, 41(7), 7–22.

Turner, K., Crook, T. R., & Miller, A. (2014). Construct measurement in social entrepreneurship: a review and assessment. En Social Entrepreneurship and Research Methods (pp. 1–18). Emerald Group Publishing Limited. https://doi.org/10.1108/S1479-838720140000009009

Villegas-Mateos, A.O., & Amorós, J.E. (2019). Regional entrepreneurial ecosystems in Mexico: A comparative analysis. Journal of Entrepreneurship in Emerging Economies, 11(4), 576–597.

Villegas-Mateos, A., & Vázquez-Maguirre, M. (2020). Social entrepreneurial ecosystems: A regional perspective of Mexico. International Journal of Entrepreneurship, 24(1), 1–19.

Yager, R.R., & Filev, D.P. (1999). Induced ordered weighted averaging operators. IEEE Transactions on Systems, Man, and Cybernetics, Part B (Cybernetics), 29(2), 141–150.

Yager, R., & Kacprzyk, J. (2012). The ordered weighted averaging operators: Theory and applications. SPRINGER, Switzerland.

5 Supply Chain Performance Assessment Using the TODIM Multicriteria Method

Toward Timely Supply

Alejandra Holguín Ávila, Luis Asunción Pérez Domínguez, Roberto Romero López, David Luviano Cruz

5.1 INTRODUCTION

The focus of this chapter is to understand the importance of having a *preemptive supply chain* in the face of potential disruptions, taking into account the radical transformations that may occur due to technological acceleration, globalization, and growing pressure for sustainability. Under this framework, where the outlook is changing, with logistical disruptions, health crises, and geopolitical conflicts, organizations require new tools to help evaluate their performance and make more agile, accurate, and resilient decisions. Given this scenario, the contemporary challenges lie in ensuring timely supplies capable of responding efficiently to complex scenarios. In this context, the use of multicriteria methods for decision-making becomes a good option, as they allow for the simultaneous integration and weighing of multiple variables.

Therefore, this chapter offers an integrated methodological approach that combines the TODIM Multicriteria method with the characteristics of Digital Twins (DTs). Through this synergy, we seek to offer a way to improve the assessment of supply chain performance and thereby strengthen its adaptive capacity in the face of uncertainty.

5.2 THEORETICAL FRAMEWORK

5.2.1 Understanding Supply Chain Performance Assessment

Starting by defining the core theme, the *Supply Chain* is a network of entities that collaborate to deliver products or services from their origin to the final customer (Kumar et al., 2019; Nakano, 2019). It encompasses diverse activities, such as *purchasing, production, logistics, and sales* (Nakano, 2019). Recent innovations in this field include *blockchain technology*, which *improves transparency and reduces communication problems* (Kumar et al., 2019), and *artificial intelligence,* which

DOI: 10.1201/9781003724902-6

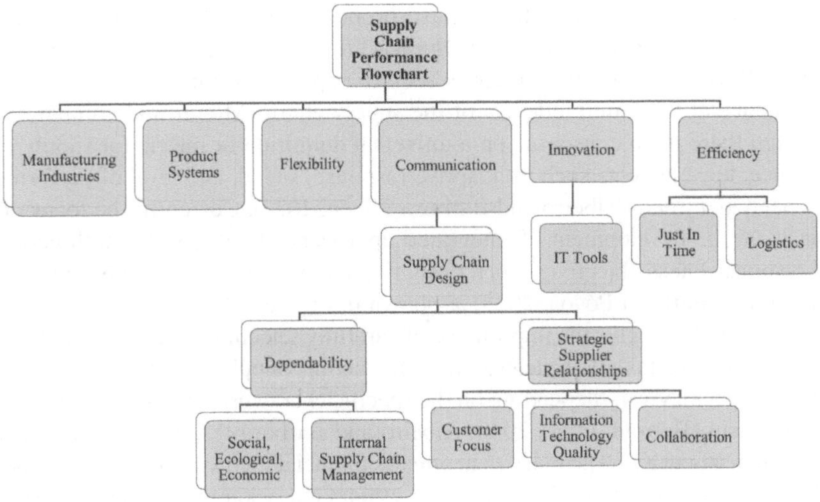

FIGURE 5.1 Supply chain performance flowchart.

improves efficiency and productivity (Charles et al., 2023). Supply chain mapping is crucial for maintaining visibility into increasingly complex networks (MacCarthy et al., 2022). Advanced techniques such as deep learning are being explored to automatically extract supply chain maps from news articles, which could aid in *risk management and network verification* (Wichmann et al., 2020).

In this way, supply chain performance refers to the activities of the extended chain to meet end customer requirements, including product availability, on-time delivery, and all inventory and capacity needed in the chain to deliver that performance responsively (Hausman, 2004). Its performance is also influenced by several *key factors* as shown in Figure 5.1: production *systems, flexibility, communication, and innovation significantly impact performance* (Mabrouk, 2020). Supply chain design, efficiency, collaboration, and information technology are crucial, with complexity and centralization being prominent factors (Sharma et al., 2024). IT tools, just-in-time manufacturing, and logistics are crucial for implementing an efficient supply chain (Panigrahi et al., 2024). Flexibility and reliability directly affect supply chain performance, while strategic supplier relationships mediate risk mitigation (Moh'd Anwer, 2023). Social, ecological, and economic factors are important for integrating Industry 4.0 and sustainability (Gopal et al., 2024). Supplier relationship management, internal supply chain management, and customer relationship management are key attributes for performance measurement (Saleheen and Habib, 2023). Customer focus, quality information technology, and collaboration among supply chain members are critical success factors (Chau et al., 2021). Lean manufacturing practices positively impact supply chain relationships and performance (Vanichchinchai, 2019).

5.2.1.1 Challenges in Supply Chain Management

Currently, the *challenges associated* with supply chain management are based on satisfying the *resilience* that it may need, which is defined as the ability to recover quickly from disruptions and likewise represents a crucial concern in modern supply chain management. Several studies have proposed metrics to quantify and model supply chain resilience (Munoz and Dunbar 2015; Behzadi et al., 2020; Ribeiro and Barbosa-Póvoa, 2023). These metrics seek to capture the multidimensional nature of resilience at the multiple levels of the supply chain (Munoz and Dunbar, 2015). Although there is no consensus on a universal definition or metric for supply chain resilience, key elements such as adaptive response, speed, and level of performance have been identified (Ribeiro and Barbosa-Póvoa, 2023). Currently, the focus of this topic is on the development of integrated approaches that consider both economic and responsiveness objectives when designing and planning resilient Supply Chains (Ribeiro and Barbosa-Póvoa, 2023) as shown in Figure 5.2.

This figure highlights the importance of carefully selecting appropriate metrics and avoiding one-size-fits-all strategies when managing Supply Chain resilience, as the best approach may vary depending on the specific characteristics and objectives of the Supply Chain (Behzadi et al., 2020; Ribeiro and Barbosa-Póvoa, 2023). Key performance indicators (KPIs) primarily encompass financial metrics, customer satisfaction, and operational efficiency (Foroughi et al., 2004; Gunasekaran and Kobu, 2007).

Environmental metrics focus on resource use and emissions, while social metrics address working conditions and community impact (Okay et al., 2024). The complexity of global operations requires advanced performance measurement systems that can integrate with emerging technologies such as the Internet of Things (IoT) (Khan et al., 2020). However, it is important to know that the challenges in implementing these systems include data reliability, information system complexity, and the need for efficient collaborative frameworks to achieve the proposed purposes (Papakiriakopoulos and Pramatari, 2010). In this way, it is important to balance metrics that have data that is difficult to collect (Abdel-Basset et al., 2019) and adapting measurement approaches to address current problems in supply chain management (Groznik and Trkman, 2009). Figure 5.3 presents a detailed visualization of these approaches in a comprehensive way for their evaluation.

Likewise, it can be said that *resilient* supply chain management demands a deep and contextualized understanding of the metrics that allow for evaluating its performance from multiple dimensions: economic, social, and environmental. The lack of a universal metric requires adopting flexible and integrated approaches capable of adapting to the complexity of each organizational environment and the demands of the global market. This overview establishes the foundation for understanding the strategic role of performance evaluation, not only as a control tool, but as a central component in decision-making aimed at the competitiveness, sustainability, and agility of the supply chain.

5.2.2 Fundamentals of the TODIM Method and Its Application

The TODIM method is a multi-decision-making approach based on prospect theory, which effectively models the psychological behavior of experts (Gomes, 2009; Ren et al., 2016). It has been extended to address various decision-making scenarios, including group decision-making (Rosli et al., 2020) and ranking problems (Wang et al., 2023). Recent developments include the generalized TODIM, which overcomes shortcomings such as weighting consistency and monotonicity (Rosli et al., 2020) and integration with other theories such as cumulative prospect theory (Zhang et al., 2022; Zhao et al., 2021).

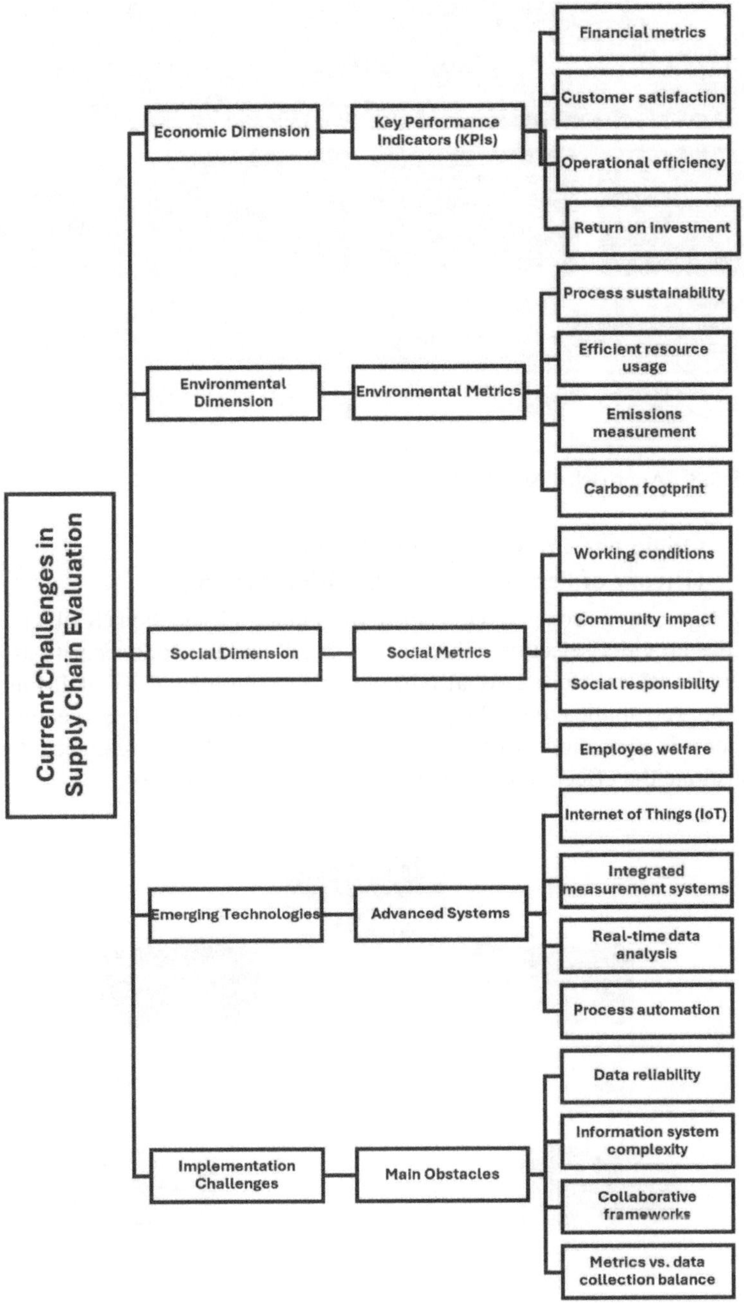

FIGURE 5.2 Critical metrics and performance measurement systems.

FIGURE 5.3 Approach to supply chain management.

5.2.2.1 Origin and Principles of the TODIM Method (Theory of Perspective and Decision-Making)

TODIM applications cover diverse fields, from material selection (Rosli et al., 2020) to logistics provider selection (Zhang et al., 2022) and network security assessment (Zhao et al., 2021). Efforts have also been made to automate the TODIM method by developing web applications (Rangel et al., 2011). An integrated perspective of these components is presented in Figure 5.4, which depicts the interconnected elements that constitute the effective implementation of this methodology.

FIGURE 5.4 TODIM process flow.

This integration can be explained as follows, based on the theory proposed by Jawad Ali and Ioan-Lucian Popa (Ali and Popa, 2025):

1. Theoretical foundation based on prospect theory: It is based on prospect theory, using its value function to represent decision-makers' subjective perceptions of profits or losses, rather than simply analyzing objective data.
2. Relative comparison process between alternatives: Comparisons are made between each alternative and the others, determining the degree of superiority of one over the other based on each criterion evaluated. From this analysis, a matrix is created that shows the relative dominance between the options.
3. Incorporating risk attitudes: One of the method's distinctive features is that it allows the analysis to be adjusted based on the evaluator's level of risk aversion or preference. This aspect is integrated into the value function through a specific parameter, allowing for the modeling of psychological tendencies and achieving decisions more aligned with the experts' actual preferences.
4. Calculating the overall dominance index: By summing up the weighted advantages that an alternative presents relative to the others, an aggregate value is obtained that represents its level of dominance. This overall index indicates how advantageous an option is compared to the rest of the evaluated set.
5. Ranking and selection of alternatives: Once the overall index for each alternative has been calculated, they are ranked based on their score. The option with the highest score is considered the most appropriate or preferred by decision-makers.
6. Extensions and adaptations: The method has been extended to adapt to contexts with imprecise or uncertain data, using representations such as trapezoidal fuzzy numbers, neutrosophic sets, or t-spherical structures. It has also been combined with techniques for assigning weights to criteria, such as the CRITIC method, which increases its versatility and applicability in different multicriteria analysis scenarios.
7. Classification of alternatives:

   Order the alternatives from highest to lowest according to their v_i values
   The alternative with the highest v_i is the most preferable.

5.2.3 AHP Analytic Hierarchy Analysis

Analytic hierarchy analysis (AHP) is a decision-making methodology that helps structure complex problems into a hierarchy of objectives, criteria, and alternatives, allowing decision-makers to make pairwise comparisons between elements, assigning priorities and weights that reflect the relative importance of each factor. The central idea is to transform subjective decisions into a quantitative process, facilitating the integration of expert judgments and historical data to obtain a coherent and consistent assessment of the available options. It is used in areas such as project management, project selection, risk assessment, and strategic planning (Cooper et al., 2006).

5.2.4 The Preference Selection Index (PSI)

The Preference Selection Index (PSI) method is a decision-making tool that considers multiple criteria at once without needing to assign specific weights to them. This makes it especially useful in complex situations, such as selecting suppliers or evaluating sustainable energy technologies (Rodríguez et al., 2024; Van Dua, 2025). Unlike other traditional methods, PSI simplifies the process by eliminating the complicated step of calculating the weight of each criterion (Van Dua, 2025). Because of this, it offers a more detailed and sensitive way to rank options, which improves accuracy when making decisions in contexts with many factors to evaluate (Sarı, 2019).

5.2.5 Digital Twins in the Supply Chain

DT technology is revolutionizing supply chain management by creating virtual replicas of physical systems, enabling real-time monitoring, predictive analytics, and optimization (Orozco-Romero et al., 2019; Odisho et al., 2024). DTs, combined with agent-based models and predictive analytics, improve decision-making, risk mitigation, and resource optimization in complex supply chains (Orozco-Romero et al., 2019; Onma Enyejo et al., 2024). By incorporating sustainable project management practices and frameworks such as the Triple Bottom Line, organizations can balance economic, social, and environmental objectives while leveraging DT technology (Onma Enyejo et al., 2024).

5.2.5.1 Architecture of a Logistics Digital Twin (DT)

Recent research highlights the importance of DTs for optimizing logistics systems. Park et al. (2020) propose a cyber-physical logistics system (CPLS) that integrates agent-based cyber-physical production systems to improve supply chain resilience in make-to-order environments. Their framework addresses problems such as the bullwhip effect and the domino effect through distributed DT simulations. Korth et al. (2018) present a modularized system architecture that combines real-time DTs with simulation logic, facilitating efficient decision-making in logistics management. This integration minimizes the need for extensive offline model maintenance and improves real-time operational capabilities. Their proposed architecture, which evolves from the digital model to the DT, aims to improve connectivity and agility while continuously adapting to changing operating conditions (Park et al., 2020; Korth et al., 2018).

5.3 METHODOLOGY

The model was applied to evaluate five suppliers in the electronics manufacturing sector. Ten criteria combining different criteria were evaluated, which are listed in Table 5.1. Each supplier was evaluated under these criteria, and the hybrid AHP–PSI–TODIM model generated the final rankings.

TABLE 5.1
Comparison of Multicriteria Methods

Details of the Evaluation Criteria

Sl. No.	Criteria	Type	Symbol	Weight
1	On time deliveries OTD (lead time)	Min	C1	0.237
2	Product quality	Max	C2	0.185
3	Operational flexibility	Max	C3	0.135
4	Total cost	Max	C4	0.114
5	Supplier reputation and reliability	Min	C5	0.101
6	Level of technological integration	Min	C6	0.075
7	Sustainability practices	Min	C7	0.057
8	Resilience to disruptions	Min	C8	0.037
9	Innovation and continuous improvement	Max	C9	0.035
10	Customer service level	Min	C10	0.024

FIGURE 5.5 Methodology application of the hybrid model.

It's worth noting that these criteria were applied to five different suppliers within the same company for an item whose value, volume, and use are high, and its availability in the market is low, making it a critical part of production.

For proper execution, this methodology must be executed as shown in Figure 5.5.

5.4 RESULTS

5.4.1 Results, Limitations, and Opportunities Detected

The results obtained from applying the TODIM method in conjunction with AHP and PSI determine the ranking of the suppliers described in alphabetical and consecutive order. Comparing the TODIM-AHP and TODIM-PSI methods, both methods agree on the positioning of supplier A5 as the best alternative, assigning a value of 1.0, and similarly, supplier A2 is ranked second in both approaches as can be seen in Table 5.2.

Furthermore, the analysis suggests a possible improvement through combining both approaches: by integrating the hierarchical logic of AHP with the objective nature of PSI, it's possible to develop a more balanced mixed model. Additionally, the TODIM method contributes further value to the analysis, which introduces an additional perspective by considering how decision-makers react to risk situations, an aspect that is especially relevant in highly uncertain environments such as the supply chain.

TABLE 5.2
Comparison of Multicriteria Methods

Suppliers	AHP-TODIM	Ranking	PSI-TODIM	Ranking
A1	0.889	3	0.594	4
A2	0.972	2	0.818	2
A3	0.797	4	0.598	3
A4	0.000	5	0.000	5
A5	1.000	1	1.000	1

5.5 DISCUSSION

5.5.1 Advantages of Traditional Methodologies

This analysis reveals an opportunity for improvement in the integration of the methods described. By combining the hierarchical structure of AHP with the objectivity of PSI, it is possible to build a more balanced hybrid model. Moreover, the inclusion of the TODIM method adds an additional layer of analysis by considering the decision-maker's behavior under risk, which is particularly relevant for examining contexts involving uncertainty and, above all, for generating more well-founded contributions.

5.5.1.1 Validation of Results

For this section, it is necessary to validate the results obtained through the application of the TODIM method combined with the AHP and PSI methods. To achieve this, a comparison was conducted between the rankings generated by both approaches. This validation was structured using a consistent set of suppliers—specifically, suppliers 1 through 5—and a homogeneous system of evaluation criteria, allowing for an analysis of consistency across methods and the identification of significant divergences, as illustrated in Figure 5.6.

The results reveal a strong alignment between both variants of the TODIM model, particularly in identifying supplier **A5** as the most favorable option, achieving a perfect score of **1.0** under both approaches. Likewise, supplier **A2** consistently ranks second, suggesting a high degree of robustness in the model's performance for these two alternatives.

Nevertheless, slight variations were observed in the ranking of suppliers **A1** and **A3**, highlighting the model's sensitivity to the weighting technique employed (AHP or PSI). These discrepancies underscore the importance of cross-validation in multicriteria decision-making environments, especially in contexts where decision-maker behavior and priority structures can significantly influence the final outcomes.

Overall, this validation supports the viability of the proposed model and suggests that integrating TODIM with complementary techniques such as AHP or PSI yields consistent, reliable, and context-adaptable results for decision-making within the supply chain.

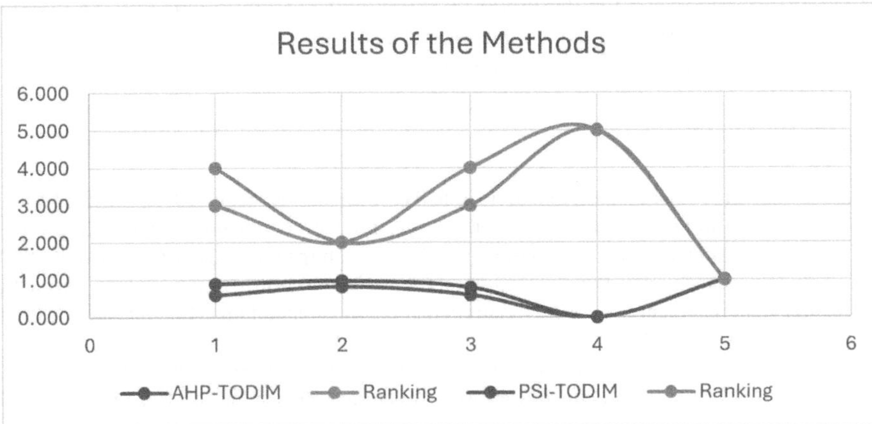

FIGURE 5.6 Results of the methods.

5.6 CONCLUSION

5.6.1 Critical Success Factors

The successful implementation of the TODIM-AHP/PSI hybrid model depends on several key factors:

- ✓ Reliable data as a foundation: For the model to deliver trustworthy results, it's crucial to work with consistent and accurate data on supplier performance based on well-defined criteria.
- ✓ Well-defined criteria and goals: It's important to clearly establish what will be evaluated and why, ensuring that all selected criteria are truly aligned with both the company's strategic procurement goals and its day-to-day operational needs.
- ✓ Involving the right people: Including key team members—such as those in procurement, logistics, and analysis—throughout the decision-making and weighting process encourages broader support and helps achieve more balanced, informed outcomes.
- ✓ Having the right tools in place: The use of appropriate digital tools—whether it's spreadsheets, custom software, or decision-support systems—can significantly ease the process of calculation, simulation, and visualization of results.
- ✓ Developing skills and understanding: More than just knowing how the model works technically, it's essential for teams to also grasp the behavioral insights that TODIM introduces, so decisions can reflect not just data but also human judgment under uncertainty.

Furthermore, despite the inclusion of behavioral elements within the TODIM framework, attention should still be given to identifying and minimizing potential biases from decision-makers during the weighting process. Finally, for the model to be sustainable over time, it should be integrated into the organization's existing procurement and supply chain management systems, ensuring seamless alignment with broader decision-support infrastructures.

REFERENCES

Abdel-Basset, M., Mohamed, R., Zaied, A. E.-N. H., & Smarandache, F. (2019). A hybrid plithogenic decision-making approach with quality function deployment for selecting supply chain sustainability metrics. Symmetry, 11(7), 903. https://doi.org/10.3390/sym11070903

Ali, Jawad, & Ioan-Lucian Popa. TODIM method with unknown weights under t-arbicular fuzzy environment for optimal gate security system selection. AIMS Mathematics, 10 (6), 13941–73 (2025).

Behzadi, Golnar, Michael Justin O'Sullivan, & Tava Lennon Olsen. On metrics for supply chain resilience. European Journal of Operational Research, 287 (1), 145–58 (2020). https://doi.org/10.1016/j.ejor.2020.04.040

Charles, Vincent, Ali Emrouznejad, & Tatiana Gherman. A critical analysis of the integration of blockchain and artificial intelligence for supply chain. Annals of Operations Research, 327 (1), 7–47 (2023). https://doi.org/10.1007/s10479-023-05169-w

Chau, Ka-Yin, Yuk Ming Tang, Xiaoyun Liu, Yun-Kit Ip, & Yiran Tao. Investigation of critical success factors for improving supply chain quality management in manufacturing. Enterprise Information Systems, 15 (10), 1418–37 (2021). https://doi.org/10.1080/17517575.2021.1880642

Cooper, William W., Lawrence M. Seiford, & Kaoru Tone. 2006. Introduction to data envelopment analysis and its uses: With DEA-solver software and references. Springer Science & Business Media.

Foroughi, Abbas, Mehmet C. Kocakulah, & William C. Perkins. Performance metrics: Key to success in global supply chain management. Journal of Transnational Management Development, 8 (3), 35–45 (2004). https://doi.org/10.1300/J130v08n03_04

Gomes, Luiz Flávio Autran Monteiro. An application of the TODIM method to the multicriteria rental evaluation of residential properties. European Journal of Operational Research, 193 (1), 204–11 (2009). https://doi.org/10.1016/j.ejor.2007.10.046

Gopal, P. R. C., Punitha Kadari, Jitesh J. Thakkar, & Bimal Kumar Mawandiya. Key performance factors for integration of industry 4.0 and sustainable supply chains: A perspective of Indian manufacturing industry. Journal of Science and Technology Policy Management, 15 (1), 93–121 (2024). https://doi.org/10.1108/JSTPM-10-2021-0151

Groznik, A., & Trkman, P. (2009). Upstream supply chain management in e-government: The case of Slovenia. Government Information Quarterly, 26(3), 459–467. https://doi.org/10.1016/j.giq.2008.12.017

Gunasekaran, Angappa, & Bulent Kobu. Performance measures and metrics in logistics and supply chain management: A review of recent literature (1995–2004) for research and applications. International Journal of Production Research, 45 (12), 2819–2840 (2007). https://doi.org/10.1080/00207540600806513

Hausman, Warren. Supply chain performance metrics. *The practice of supply chain management: Where theory and application converge*, 61–73 (2004).

Khan, Sharfuddin Ahmed, Amin Chaabane, & Fikri Dweiri. Supply chain performance measurement systems: A qualitative review and proposed conceptual framework. International Journal of Industrial and Systems Engineering, 34 (1), 43–64 (2020). https://doi.org/10.1504/IJISE.2020.104315

Korth, B., Schwede, C., & Zajac, M. (2018). Simulation-ready digital twin for realtime management of logistics systems. 2018 IEEE International Conference on Big Data (Big Data), 4194–4201. https://doi.org/10.1109/BigData.2018.8622160

Kumar, Kaushik, Divya Zindani, & J. Paulo Davim. 2019. Industry 4.0: developments towards the fourth industrial revolution. Springer.

Mabrouk, Nejah Ben. Analysis of critical factors influencing the supply chain performance of small and medium enterprises. Journal of Advanced Manufacturing Systems, 19 (4), 781–97 (2020). https://doi.org/10.1142/S0219686720500328

MacCarthy, Bart L., Wafaa A. H. Ahmed, & Güven Demirel. 2022. Mapping the supply chain: Why, what and how?. International Journal of Production Economics (2022).

Moh'd Anwer, AL-Shboul. Enabling manufacturing firms' supply chain performance in the Middle East region through boosting the quality of multi-directional relationship, and supply chain risk dilution: A moderated-mediation model. Heliyon 9 (11) (2023). https://doi-org.psut.idm.oclc.org/10.1108/978-1-80071-897-520221018

Munoz, Albert, & Michelle Dunbar. On the quantification of operational supply chain resilience. International Journal of Production Research, 53 (22), 6736–51 (2015). https://doi.org/10.1080/00207543.2015.1057296

Nakano, Mikihisa. What Is a Supply Chain? *Supply Chain Management* (2019). https://api.semanticscholar.org/CorpusID:201441482

Odisho, J., Karamchandani, M., Swadley, E., Awwad, M., & Püttner, M. (2024, mayo 9). Supply Chain Digital Twins: A Comprehensive Review and Case Studies. Proceedings of the International Conference on Industrial Engineering and Operations Management. 5th South American International Conference on Industrial Engineering and Operations Management. https://doi.org/10.46254/SA05.20240069

Okay, Nilufer C., Asli Sencer, & Nazim Taskin. Quantitative indicators for environmental and social sustainability performance assessment of the supply chain. Environment, Development and Sustainability, 1–21 (2024). https://doi.org/10.1007/s10668-024-05210-3

Onma Enyejo, J., Peter Fajana, O., Sele Jok, I., Judith Ihejirika, C., Olusola Awotiwon, B., & Motilola Olola, T. (2024). Digital Twin Technology, Predictive Analytics, and Sustainable Project Management in Global Supply Chains for Risk Mitigation, Optimization, and Carbon Footprint Reduction through Green Initiatives. International Journal of Innovative Science and Research Technology (IJISRT), 609–630. https://doi.org/10.38124/ijisrt/IJISRT24NOV1344

Orozco-Romero, A., Arias-Portela, C. Y., & Saucedo, J. A. M.-. (2020). The Use of Agent-Based Models Boosted by Digital Twins in the Supply Chain: A Literature Review. En P. Vasant, I. Zelinka, & G.-W. Weber (Eds.), *Intelligent Computing and Optimization* (Vol. 1072, pp. 642–652). Springer International Publishing. https://doi.org/10.1007/978-3-030-33585-4_62

Panigrahi, Swayam Sampurna, Rajesh Katiyar, & Debasish Mishra. Integrated DEMATEL-ML approach for implementing lean supply chain in manufacturing sector. Journal of Advances in Management Research, 21 (3), 449–77 (2024). https://doi.org/10.1108/JAMR-08-2023-0231

Papakiriakopoulos, D., & Pramatari, K. (2010). Collaborative performance measurement in supply chain. Industrial Management & Data Systems, 110(9), 1297–1318. https://doi.org/10.1108/02635571011087400

Park, K. T., Son, Y. H., & Noh, S. D. (2020). The architectural framework of a cyber physical logistics system for digital-twin-based supply chain control. International Journal of Production Research, 59(19), 5721–5742. https://doi.org/10.1080/00207543.2020.1788738

Rangel, Luís Alberto Duncan, Luiz Flavio Autran Monteiro Gomes, & Felipe Parussoli Cardoso. An application of the TODIM method to the evaluation of broadband internet plans. Pesquisa Operacional, 31, 235–49 (2011). https://doi.org/10.1590/S0101-74382011000200003

Ren, Peijia, Zeshui Xu, & Xunjie Gou. Pythagorean fuzzy TODIM approach to multi-criteria decision making. Applied Soft Computing, 42, 246–59 (2016). https://doi.org/10.1016/j.asoc.2015.12.020

Ribeiro, João Pires, & Ana Paula FD Barbosa-Póvoa. A responsiveness metric for the design and planning of resilient supply chains. Annals of Operations Research, 324 (1), 1129–81 (2023). https://doi.org/10.1007/s10479-022-04521-w.

Rodríguez, Ángela Rodríguez, Raquel Irene Laguado Ramírez, Belisario Peña Rodríguez, & Lida Yaneth Maldonado Mateus. 2024. Selección de proveedores con el método PSI. Mundo FESC, advance online publication. https://doi.org/10.61799/2216-0388.1668

Rosli, NNNC, Binyamin Yusoff, CMIC Taib, & Elissa Nadia Madi. TODIM under group decision making based on weighted selective aggregated majority-OWA operator. International Journal of Engineering Trends and Technology, 134–42 (2020).

Saleheen, Ferdoush, & Md Mamun Habib. "Embedding attributes towards the supply chain performance measurement". Cleaner Logistics and Supply Chain 6, 100090 (2023). https://doi.org/10.1016/j.clscn.2022.100090

Sarı, Emre Bilgin. Measuring the performances of the machines via Preference Selection Index (PSI) method and comparing them with values of Overall Equipment Efficiency (OEE). İzmir iktisat dergisi 34 (4), 573–81 (2019). https://doi.org/10.24988/ije.2019344859

Sharma, Rohit, Anjali Shishodia, Sachin Kamble, Angappa Gunasekaran, & Amine Belhadi. Agriculture supply chain risks and COVID-19: Mitigation strategies and implications for the practitioners. International Journal of Logistics Research and Applications, 27 (11), 2351–77 (2024). https://doi.org/10.1080/13675567.2020.1830049

Van Dua, Tran. Application of the PSI method in selecting sustainable energy development technologies. Engineering, Technology & Applied Science Research, 15 (1), 19596–601 (2025). https://doi.org/10.48084/etasr.9317

Vanichchinchai, Assadej. The effect of lean manufacturing on a supply chain relationship and performance. Sustainability, 11 (20), 5751 (2019).

Wang, Liang, Zi-Xin Zhang, Alessio Ishizaka, Ying-Ming Wang, & Luis Martínez. TODIMSort: A TODIM based method for sorting problems. Omega, 115, 102771 (2023). https://doi.org/10.1016/j.omega.2022.102771

Wichmann, Pascal, Alexandra Brintrup, Simon Baker, Philip Woodall, & Duncan McFarlane. Extracting supply chain maps from news articles using deep neural networks. International Journal of Production Research, 58 (17), 5320–36 (2020). https://doi.org/10.1080/00207543.2020.1720925

Zhang, Dan, Yong Su, Mengwei Zhao, & Xudong Chen. CPT-TODIM method for interval neutrosophic MAGDM and its application to third-party logistics service providers selection. Technological and Economic Development of Economy 28 (1), 201–19 (2022).

Zhao, Mengwei, Guiwu Wei, Jiang Wu, Yanfeng Guo, & Cun Wei. TODIM method for multiple attribute group decision making based on cumulative prospect theory with 2-tuple linguistic neutrosophic sets. International Journal of Intelligent Systems, 36 (3), 1199–222 (2021). https://doi.org/10.1002/int.22338

6 Assessment of Student Satisfaction in Upper Secondary Education

A Strategy for Continuous Improvement

Lydia María López-Barraza,
Diego Alonso Gastélum-Chavira,
Alanda Catalina Torres-Salazar, and
María Guadalupe López-Barraza

6.1 INTRODUCTION

The student satisfaction is a key indicator of educational quality in scholar institutions. A clear example is in university contexts where it is used to identify the extent to which the student's expectations and needs are met during their transition. This assessment provides a tool for decision-making by managers to generate improvements in services, as well as in the association that the greater student satisfaction, the greater academic commitment, motivation and school permanence (Garbanzo Vargas, 2012; Surdez-Pérez et al., 2018).

It is necessary to find reliable ways to measure it so that institutions know their reality, can compare it with other competitors and analyze it over time (Cadena-Badilla et al., 2015). Here the role of students is fundamental, since the assessment of institutional conditions and substantive functions can generate improvement in the quality of services, positively impacting evaluations and accreditations carried out by external organizations (Marcano Molano & Uribe Veintimilla, 2022).

And even when institutions try to carry out the evaluation, the lack of a comprehensive approach and the scarce existing methodology may allow important aspects of school life to be neglected (Haverila et al., 2021; Lozano Treviño & Rodríguez Rodríguez, 2024), Which could affect its impact as a feedback tool.

The data collection instruments for evaluation must be designed in consideration of the context, the pedagogical model, and the institutional programs. Which enables the making of an accurate diagnosis, facilitating the implementation of strategies tailored to the educational conditions of the organization (Surdez-Pérez et al., 2018).

This chapter analyzes the results of a satisfaction evaluation implemented in a Mexican system of Upper Secondary Education (USE), this school grade attends

DOI: 10.1201/9781003724902-7

young people between the ages of 15 and 18, and according to UNESCO International Standard Classification of Education corresponds to level 3, which encompasses programs designed to complement secondary education and prepare students for tertiary education (university studies) or the labor market (UNESCO Institute for Statistics, s. f.).

For the approach of the evaluation tool studied, the dimensions and indicators were identified according to the central activities of the management areas responsible for its operation, academic and administrative. After its application, a detailed analysis of each indicator was executed, allowing those with lower levels of satisfaction to be identified. This permitted the development of specific actions aimed at improving the experience of students throughout their academic journey.

The document is organized initially with a conceptual theoretical review. Subsequently, the description of the methodology is presented, a results section, as well as a space for the presentation of proposals for continuous improvement. And finally, the conclusion.

6.2 THEORETICAL AND CONCEPTUAL FRAMEWORK

There is not a single definition of educational quality, as it largely depends on each specific contexts of the institutions (Egido, 2005; Vázquez, 2013). Therefore, its assessment requires the delineation of dimensions and indicators aligned with their existing realities (Surdez-Pérez et al., 2018). Nonetheless, specific parameters have been identified by international organizations such as the United Nations Educational, Scientific and Cultural Organization (UNESCO, 2014), which recommends guaranteeing quality education for all one, that fosters a permanent learning. Among these factors stand out: well trained and motivated teachers; offer relevant content that enables students to acquire sufficient knowledge and the competencies appropriate to their academic degree; provide safe and inclusive learning environments; contain programs for the development of students' capacities to become creative and innovative individuals capable of adapting to social changes and work demands.

Other authors add to these parameters the following (Daghlian, 1997; Lema, 2007; Martínez et at., 2006; Panera, 1999; Sánchez, 2008): place the teachers as the central axis to encourage educational quality; the promotion of policies for the continuous improvement of teachers in training as updating; encourage innovation in teaching-learning methods in the classroom; and have an organizational structure to support the general well-being of the school community.

Additional aspects to consider include an adequate budgeting and infrastructure (Jiménez et al., 2011; Lema, 2007); the periodic implementation of user satisfaction evaluation (Bartolomé et al., 2006; Bogue & Saunders, 1992; Maldonado et al., 2013); the promotion of self-management and self-learning of students, allowing them to define and advance in their life project (Astin & Antonio, 2012; Pérez, 2004); the offer of programs aligned with the mission and values of the institution that contribute significantly to students' lives (Bergquist, 1995); and the relevance of having reports to the social context on which it focuses (Braslavsky, 2006; Clemenza et al., 2005).

Regardless of the parameters identified and those selected by each institution, it is important to mention that additional study have highlighted the benefit of teachers having the knowledge and being aware of the variables that influence student satisfaction to ensure an optimal experience (Fernández-Rodicio & Abellán-Roselló, 2024).

On the other hand, focusing on this evaluation as a tool for assessing educational quality, it is from the perspective of the students that it must be implemented as beneficiaries of the services (Cadena-Badilla et al., 2015; González, 2006; Jiménez et al., 2011; Salinas et al., 2008; Sandoval, 2006; Vázquez, 2013; Vergara & Quesada, 2011).

At this point and through the literature analyzed, it can be mentioned that student satisfaction is understood as the feeling of well-being that users experience when their academic and personal expectations are met, through the training actions of the institution and its complementary services.

Until the conclusion of this research, no studies had been identified that specifically addressed student satisfaction in USE. However, it was possible to find diverse publications focused on Higher Education, where it was possible to identify the proposal of indicators to evaluate educational quality and its implications for academic life (Alves & Raposo, 2005; Caballero et al., 2007; Garbanzo Vargas, 2012; Osorio & Pérez, 2010; Sinclaire, 2014; Surdez-Pérez et al., 2018).

These articles have shown that educational quality improves the reputation and prestige of a school organization. In addition, several dimensions are identified to measure student satisfaction, such as staff support, empathy generated for the basic needs of the community, complementary services, the academic offer of the study plans, the management of its directors, individual attention and the capacity for institutional response to problems (Mejías & Martínez, 2009; Mejías et al., 2006). To which are attached vital and emotional security, equal opportunities to develop, a sense of recognition with the instance, the presence of clear teaching systems, the impulse to personal development, success and the satisfaction of self-realization (Gento & Vivas, 2003).

In conclusion, this type of evaluation allows to identify areas for continuous improvement to satisfy the academic and personal development needs of students, highlighting educational institutions in an inclusive training space. The challenge is to propose measurement instruments in response to the contexts, generate assessments in accordance with its organizational system, academic offer and services.

6.3 METHODOLOGY

The study adopted a quantitative methodology, since it focuses on the collection and analysis of numerical data that were obtained through the application of a structured instrument intended to students. This approach allowed for an objective measurement of student satisfaction levels across several predefined dimensions, as well as the identification of patterns, trends, and areas for improvement.

The design of the instrument began with qualitative analysis focused on documentary review and interviews with institutional managers, enriching the information for the definition of the dimensions to be evaluated. In such a way that mixed methods were integrated throughout the research. However, the core of the analysis and interpretation of the results is based on quantitative methods, which is why this is considered the predominant approach.

The development of the student satisfaction evaluation instrument followed a structured methodology consisted of the following phases:

1. Document review: The Institutional Development Plan (PDI) of the organization was analyzed to identify strategic guidelines, priority objectives and conducting principles of its educational management. In addition, the basic academic institutional programs of the different areas (academia, culture, sports, educational orientation) were reviewed to understand their scope, structure and implementation criteria.
2. Interviews with key stakeholders: Semi-structured interviews were carried out with the management areas responsible for implementing the central programs, monitoring the PDI and compliance with the programmed goals. These interviews provided direct information about the interests, needs, and expectations of information to be obtained from the evaluation of student satisfaction.
3. Identification of priority topics: Based on the documentary analysis and the results of the interviews, it was possible to identify the central themes to classify them by areas of interest, guiding the evaluation to a structure that would allow defining strategies for continuous institutional improvement.
4. Definition and operationalization of dimensions: The dimensions to be evaluated were established, considering both institutional priorities and the context. Subsequently, the indicators for measuring the dimensions were defined and the survey questions were posed.
5. Design and validation of the instrument: Based on the previous steps, a survey was designed that was revised by two experts in academic management to ensure its relevance, clarity and coherence with the objectives of the evaluation. This stage ensured that the instrument was well suited to the institutional context and could generate valid diagnostic results.

The final product was a structured measurement tool composed of 25 indicators distributed across four dimensions, as presented in Table 6.1. An ordinal Likert scale was used (very satisfied, satisfied, somewhat satisfied and not satisfied), including the option not applicable for indicators that cannot be evaluated in some educational establishments, especially those related to infrastructure.

Questions related to institutional identity were also included with a yes or no response option. This allowed a global evaluation of student satisfaction (review Table 6.2).

Concerning the USE institution selected for the study, it is a public organization with presence in the 20 municipalities of Sinaloa in Mexico. With a total of 128 educational campuses and the capacity to enroll 25% of the state's target population, which makes it in the second-largest institution in this region with enrolled young people between the ages of 15 and 18.

TABLE 6.1
Dimensions and Indicators

Dimension	Operational Definition	Indicator
Formative components	Measure the student satisfaction with the elements that make up the teaching-learning process, including pedagogical resources, the guidance received and the educational experience.	• Quality of teaching • Acquired knowledge • Institutional textbooks • Digital educational platform • Academic advising • Educational guidance • Overall student experience • Number of students per group
Cultural and sports activities	Evaluation of the offer, access and conditions of sports and cultural spaces, to facilitate participation in extracurricular activities that impact the comprehensive development of the student body.	• Offer of cultural and sports activities • Practice spaces and equipment
Training, services and support programs	Level of satisfaction with institutional services, the training offered to complement academic training, attention to academic, administrative and personal needs.	• Complementary academic courses, workshops, and seminars • Attention from school authorities to academic needs • Support from administrative staff • Scholarship program • Student mentoring program • Security of school facilities
Infrastructure	Students' perception of the availability, condition and functionality of physical spaces and material resources such as campus equipment.	• Classrooms • Computer center • Experimental science laboratories • Job training workshops • Restrooms • Library • Auditorium • Green areas • Cafeterias

TABLE 6.2
Questions about the Institutional Identity

Question
Would you enroll in the institution again?
Would you recommend the institution to friends, family or acquaintances?
Do you feel like you are part of the institution?

This educational modality is one of the main public institutions within the USE system in Mexico, which is affiliated to the Secretariat of Public Education of the National Government. It offers a general high school modality with a duration of three years, with the objective of providing an all-inclusive education that combines a solid academic background, the specialization of work abilities and the development of life competencies. It is a stage preceding to university studies. Among its main characteristics are:

1. It is part of the centralized USE system.
2. It is aligned with the Common Curriculum Framework for USE, which promotes the development of generic competencies such as critical thinking, communication, teamwork and problem solving; as well as disciplinary, which are knowledge, skills and attitudes of each field of knowledge; and professionals to provide talents to perform in the workplace.
3. The curriculum is divided into three components:
   - The basic common core includes subjects such as mathematics, language and communication, social sciences, experimental sciences and humanities.
   - Work training program that provides students with basic technical skills to enter the labor market, in professions related to entrepreneurship, communication, accounting, nursing, information technologies, among others.
   - Strategic plans for personal and social development through classes on educational and vocational guidance, the promotion of arts and sports.
4. In addition to what is specified in the curriculum, extracurricular activities focused on academics, culture, sports and community service are carried out.
5. Education is in-person and is provided in a curricular plan of three school grades divided into six semesters.
6. Graduates receive a certificate of completion of studies with which they can apply for admission to higher education institutions in Mexico and other countries.

Other services promoted by the institution include tutoring programs, the use of digital technologies and commitment to the social environment, aligning with the principles of equity, inclusion and continuous improvement of the national educational system.

To apply the instrument, an invitation to participate was sent to all students. As a result, a total of 11,078 responses were obtained, which represent 35% of the total student population ($N \approx 31{,}651$). The study achieves a margin of error of ±0.99% with a confidence level of 99%, confirming the statistical precision and representativeness of the sample analyzed. The distribution of answers by school grade was as follows:

- 35.5% of first year students,
- 30% second year students, and
- 34.5% third year students.

The results provide a solid and representative basis for the analysis of the information obtained, strengthening the validity of the findings and their usefulness for making institutional decisions aimed at continuous improvement.

6.4 RESULTS

One of the objectives set out in the PDI has reference to the level of school satisfaction of the students. The stated goal stipulates achieving a high level of satisfaction in 90% of the students surveyed. Based on this reference, the analysis of the results indicates that on average more than half of the respondents report acceptable levels of satisfaction in all dimensions and indicators evaluated.

A particularly notable finding is related to responses to questions about institutional identity, where 94% of students reported a strong sense of belonging. This data aligns with the general perception of student satisfaction, since 82% of respondents indicated that they are very satisfied or satisfied. Furthermore, 92% would recommend the institution and 91% would enroll again.

However, it is essential to note that none of the indicators reached the projected benchmark of 90%, even when the very satisfied and satisfied response levels were combined.

For the analysis of the data obtained, the two categories of responses corresponding to very satisfied and satisfied were grouped to represent a high level of satisfaction, while all other responses were classified as medium to low satisfaction, which will facilitate interpretation and comparison in subsequent examinations.

The results of the dimension of the formative components generally showed that students have a positive perception of the evaluated indicators (see Table 6.3). The three with the highest levels of satisfaction were the overall experience of the students (82%), the acquired knowledge (82%) and the quality of teaching (81%).

These data positively valued the quality of the educational process, the support provided by teachers and their general academic experience.

Subsequent, with slightly lower but acceptable percentages, were educational guidance (75%), institutional textbooks (74%) and class size, number of students

TABLE 6.3
Student Satisfaction Rates for Formative Components

Indicators	High Satisfaction	Medium to Low Satisfaction
Overall student experience	82%	18%
Acquired knowledge	82%	18%
Quality of teaching	81%	19%
Educational guidance	75%	25%
Institutional textbooks	74%	26%
Number of students per group	74%	26%
Academic advising	69%	31%
Digital educational platform	66%	34%

TABLE 6.4
Student Satisfaction Rates for Cultural and Sports Activities

Indicators	High Satisfaction	Medium to Low Satisfaction
Offer of cultural and sports activities	66%	34%
Practice spaces and equipment	63%	37%

per group (74%). These aspects reflect an intermediate level of satisfaction, possibly linked to structural or logistical conditions that could limit a better evaluation.

Finally, the indicators with the lowest levels of satisfaction were the academic advising (69%) and the digital educational platform (66%). This may reflect limitations in personalized academic support or challenges in the use and access of technological tools. These findings suggest the need to review educational support and digital integration strategies to strengthen the learning experience.

Regarding the dimension of cultural and sports activities, the perception of the students was at a moderate percentage, with indicators below 70% as presented in Table 6.4. These areas are positioned as priorities for improvement as set out below:

- The indicator of the offer of cultural and sports activities (66%) showed that two out of three students were satisfied. However, the remaining 34% reported medium or low satisfaction, which suggests disappointment with the type of sports and cultural practices proposed by the campus, their frequency or relevance.
- The practice spaces and equipment (63%) was the lowest rated indicator within the dimension, signifying that more than a third of students considered that the spaces intended for practices such as training, as well as the equipment used to carry them out, were inadequate, insufficient or not functional. This perception may be related to physical space limitations, deficient maintenance, or restricted availability for use of the facilities.

The information highlights the need to strengthen the availability of physical spaces to have a constant agenda of cultural, artistic, sports and recreational activities to promote student well-being.

About the dimension of training, services and support programs, the responses revealed a varied perception of the different institutional services that complement the academic experience. Although some indicators reached acceptable levels of satisfaction, others are critical areas for immediate attention (see Table 6.5).

Accordingly, the following observations can be made:

- The support from administrative staff (72%), attention from school authorities to academic needs (71%), and complementary courses, workshops and seminars (69%) fell within a relatively high satisfaction range. Suggesting that students positively value the academic and administrative support received throughout their educational career.

TABLE 6.5
Student Satisfaction Rates for Training, Services and Support Programs

Indicators	High Satisfaction	Medium to Low Satisfaction
Support from administrative staff	72%	28%
Attention from school authorities to academic needs	71%	29%
Complementary courses, workshops, and seminars	69%	31%
Security of school facilities	68%	32%
Student mentoring program	61%	39%
Scholarship program	53%	47%

- The security of school facilities was perceived relatively favorably, with 68% of responses. However, the remaining 32% expressed concerns related to the maintenance of the space.
- The student mentoring program was qualified as moderately effective (61%), possibly due to lack of clarity about its operation, limited visibility or inconsistent implementation.
- The indicator with the lowest level of satisfaction was the scholarship program, with 47% of respondents expressing dissatisfaction or lack of knowledge. This is a critical alert about the program's coverage, accessibility or institutional communication, which may impact the most vulnerable student populations.

Together, these findings indicate that while there are strengths in support services and supplemental academic training, there are significant gaps in key scholar support programs, particularly in scholarships and student mentoring, that must be attended to ensure equity and student retention.

In the infrastructure dimension, responses revealed a mixed perception (see Table 6.6). While some places received acceptable satisfaction levels, several indicators reflected negative opinions that may impact the quality of the educational experience and student well-being.

According to Table 6.6, the best rated were the computer center (73%), experimental science laboratories (66%), and the library (64%), indicating that students recognized their functionality and academic value. A second group of spaces with moderate satisfaction includes classrooms (62%), job training workshops (62%) and green areas (57%), showing a divided perception between those who found these areas adequate and who perceived deficiencies. This highlights the need to address issues related to furniture, ventilation, lighting or availability of use.

The worst evaluated indicators were the restrooms (39% satisfaction), auditorium (49%) and cafeteria (44%), which obtained more than 50% of responses that are in the range of medium to low satisfaction. These results exposed a critical perception among students toward places that directly affect their comfort, hygiene, social interaction, which can negatively impact their sense of belonging.

TABLE 6.6
Student Satisfaction Rates for Infrastructure

Indicators	High Satisfaction	Medium to Low Satisfaction
Computer center	73%	27%
Experimental science laboratories	66%	34%
Library	64%	38%
Classrooms	62%	38%
Job training workshops	62%	38%
Green areas	57%	43%
Auditorium	49%	51%
Cafeteria	44%	56%
Restrooms	39%	61%

In summary, the infrastructure dimension reveals areas of technological and academic strength, but also significant weaknesses in basic services such as community spaces. Therefore, a permanent maintenance plan is necessary to guarantee decent, safe and functional conditions during the students' stay at the institution.

6.5 PROPOSAL FOR CONTINUOUS IMPROVEMENT

The USE represents a vital stage in the academic and personal development of young people between the ages of 15 and 18 in Mexico, for this reason ensuring quality is a priority for educational institutions. In this way, the following proposals aim to establish a continuous improvement plan based on the results of the student satisfaction indicators, to influence academic quality, strengthen the educational experience and promote a culture of welfare.

The vision proposed not only addresses the current needs of students but also helps to consolidate efficient, participatory educational management, oriented to innovation. To do this, the strategies focus on indicators classified as medium to low levels of satisfaction (see Table 6.7).

Based on the results obtained, it was identified that the educational digital platform has medium to low levels of satisfaction. This situation requires the design and implementation of actions aimed at improving its functionality, accessibility and pedagogical usefulness, since it constitutes an important resource for learning, academic management, and teacher-student interaction.

An area of betterment focuses on the pedagogical use of the academic platform; therefore, a permanent teacher training program must be applied with the purpose of generating interactive content, allowing the appropriate use of digital tools inside and outside the classroom, asynchronously and synchronously.

In addition, it is recommended to develop tutorials and interactive guides that facilitate their use along with a communication plan to socialize these support tools.

These actions constitute a cyclical process of continuous improvement that begins with diagnosis, followed by planning, implementation, evaluation, and feedback, thus allowing the educational platform to evolve in favor of academic quality and the student experience.

TABLE 6.7
Selected Indicators for Implementing Continuous Improvement Strategies

Indicator	Continuous Improvement Strategies
Digital educational platform	• Implement training programs to optimize platform use. • Improve and update content to ensure its relevance and quality. • Develop a communication plan to facilitate efficient access and use. • Evaluate user access, utilization, and retention every month.
Scholarship program	• Develop and circulate a clear and accessible call for applications. • Expand coverage. • Automate the application and renewal process for greater efficiency. • Prioritize the inclusion of students in situations of financial vulnerability.
Practice spaces and equipment for cultural and sports activities	• Design a space renovation program, including a service schedule and cost projection. • Address material needs for sports, civic, cultural, and artistic activities. • Implement a training program for instructors and coaches.
Restroom	• Renovate facilities in poor condition. • Ensure that people with disabilities have appropriate access. • Implement a reporting system for maintenance issues. • Assign dedicated supervisory and cleaning staff. • Promote a culture of care and respect for facilities.

The review revealed that, although the scholarship program offers essential support to many learners, the outcome is relatively modest, which is related to several factors: unclear information about requirements and procedures, limitations in beneficiary coverage, and lengthy response times. Then, it is indispensable to reinforce its operation, transparency, and accessibility.

The proposal centers on redesigning the socialization-orientation methods to respond to the call for scholarships, clearly exposing requirements, dates and the phases of the selection process. In addition, a clear, multichannel and continuous communication program must be executed, which includes promotion of the use of institutional social networks and physical information spaces.

It is suggested to expand the coverage of the scholarship program to increase the number of beneficiaries. Especially those who face economic, social vulnerability or health limitations. The institution requires a diagnosis to accurately identify those who are in situations of priority attention.

The process of requesting, reviewing and assigning scholarships needs automation through a digital platform that reduces response times, avoids duplication and facilitates consultations by applicants. The impact can be measured with indicators on the reduction of service times, decrease of administrative errors and number of beneficiaries attended.

The comprehensive development of students requires, in addition to a solid academic base, the raise of appropriate spaces for cultural and sports activities. However, satisfaction results indicate a medium to low level in this dimension, suggesting deficiencies in infrastructure, availability and quality of equipment. This situation limits opportunities for recreation, strengthening socio-emotional skills, institutional identity and physical plus mental well-being of students.

In this context, it is proposed to rehabilitate and recover the physical spaces intended for extracurricular activities, which include sports fields, classrooms for cultural workshops, common areas and auditoriums. First, an initial technical assessment is planned to identify the structural conditions, the situation of the furniture and the appropriate use.

Another action is reinforcement the equipment and materials necessary for the development of cultural and sports activities. An update of the institutional inventory is recommended, along with the acquisition of musical instruments, sports equipment, audiovisual materials, and specialized furniture, defined according to demand by activity and workshop planning.

In parallel, a preventive-corrective maintenance program must be established to ensure the conservation of spaces, define the appropriate use of equipment, carry out periodic inspections and define mechanisms to report incidents. Which should also cover the care of restrooms, another of the worst valued indicators.

These actions request to strengthen the sense of belonging, prevent risk behaviors and promote healthy lifestyles aligned with a comprehensive educational vision.

6.6 CONCLUSION

The analysis emphasizes the importance of incorporating satisfaction evaluation as a systematic, continuous and mandatory process in USE institutions. From the student's perspective, this tool drives innovation, continuous improvement, adapting the educational system to current challenges, allowing school leadership teams to respond effectively to the demands facing the educational system and the labor market.

As institutions committed to quality, educational organizations must ensure that provide adequate learning environments, comply with current regulations, maintain transparency and accountability and, above all, base their decisions on solid evidence. By doing so, they strengthen public confidence in education and help consolidate a management model oriented to the well-being of students, contributing to the comprehensive development of the school community.

For this reason, it must continue studying different dimensions and assessment indicators of school satisfaction, with the objective of generating a greater range of analysis possibilities, developing proposals that support academic organizations to encourage a culture of constant evaluation that allows them to make decisions based on reliable data.

ACKNOWLEDGMENTS

This research was financed by the Coordinación General para el Fomento a la Investigación Científica y Tecnológica del Estado de Sinaloa (CONFÍE) through the Fondo Estatal de Apoyo a la Ciencia, Tecnología e Innovación de Sinaloa (FEACTIS). The authors gratefully acknowledge this support.

REFERENCES

Alves H. & Raposo M. (2005). La medición de la satisfacción en la enseñanza universitaria: El ejemplo de la Universidad de da Beira Interior. Universidad de da Beira Interior. *International Review on Public and Nonprofit Marketing*. 1(1), 73–88.

Astin, A. W. & Antonio, L. A. (2012). *Assessment for Excellence. The Philosophy and Practice of Assessment and Evaluation in Higher Education* (2 ed.). Maryland, USA.: Rowman & Litlefield.

Bartolomé, A., Blanco, Ó., Hernández, R. & Murguey, V. (2006). La calidad en la enseñanza universitaria (una experiencia institucional). *Geoenseñanza*, 11(2), 163–173.

Bergquist, W. H. (1995). *Quality through Access, Access with Quality: The New Imperative for Higher Education*. San Francisco, CA: Jossey-Bass.

Bogue, E. G. & Saunders, R. L. (1992). *The evidence for quality: Strengthening the tests of academic and administrative effectiveness*. San Francisco, CA: Jossey-Bass.

Braslavsky, C. (2006). Diez factores para una educación de calidad para todos en el siglo XXI. *Reice-Revista Electrónica Iberoamericana sobre Calidad, Eficiencia y Cambio en Educación*, 4(2), 84–101.

Caballero, C. C., Abello, R. & Palacios, J. (2007). Relación del burnout y el rendimiento académico con la satisfacción frente a los estudios en estudiantes universitarios. *Avances en Psicología Latinoamericana*, 25(2), 98–111.

Cadena-Badilla, M., Mejías-Acosta, A., Vega-Robles, A. & Vásquez-Quiroga, J. (2015). La satisfacción estudiantil universitaria: análisis estratégico a partir del análisis de factores. *Revista Industrial Data*, 18(1), 9–18. https://www.redalyc.org/pdf/816/81642256002.pdf

Clemenza, C., Ferrer, J. & Pelekais, C. (2005). La calidad como elemento competitivo en las Universidades. *Revista de Artes y Humanidades UNICA*, 6(14), 55–83.

Daghlian, J. (1997) Educación para la calidad y calidad para la educación. *Pharos*, 4(2), 111–116.

Egido, I. (2005). Reflexiones en torno a la evaluación de la calidad educativa. *Tendencias Pedagógicas*, 10, 18–28.

Fernández-Rodicio, C.I., & Abellán-Roselló, L. (2024). Revisión sistemática de investigaciones sobre la satisfacción con los estudios universitarios. *EDUCA Revista Internacional para la Calidad Educativa*, 4(2), 383–408.

Garbanzo Vargas, G. M. (2012). Factores asociados al rendimiento académico en estudiantes universitarios, una reflexión desde la calidad de la educación superior pública. *Revista Educación*, 31(1), 43–63. https://doi.org/10.15517/revedu.v31i1.1252

Gento, S. & Vivas, M. (2003). El SEUE: un instrumento para conocer la satisfacción de los estudiantes universitarios con su educación. *Acción Pedagógica*, 12(2), 16–27.

González, I. (2006). Dimensiones de evaluación de la calidad universitaria en el espacio europeo de educación superior. *Revista Electrónica de Investigación Psicoeducativa*, 10, 4(3), 445–468.

Haverila, M., Haverila, K., McLaughlin, C., & Arora, M. (2021). *Towards a comprehensive student satisfaction model. The International Journal of Management Education*, *19*(3), 100558. https://doi.org/10.1016/j.ijme.2021.100558

Jiménez, A., Terriquez, B. & Robles, F. J. (2011). Evaluación de la satisfacción académica de los estudiantes de la Universidad Autónoma de Nayarit. *Revista Fuente*, 3(6), 46–56.

Lema, J. (2007). La calidad educativa, un tema controvertido. *Reencuentro*, 50, 10–14.

Lozano Treviño, D. F., & Rodríguez Rodríguez, D. C. (2024). Factores asociados con la satisfacción del estudiantado de bachillerato en un colegio militarizado. RIDE *Revista Iberoamericana para la Investigación y el Desarrollo Educativo*, 14(28). https://doi.org/10.23913/ride.v14i28.1940

Maldonado, S. E., Guillén, A. M., Carranza, R. E. & Ramírez, M. A. (2013). Análisis de la calidad de los servicios recibidos en las instituciones de educación superior. *Revista Internacional Administración & Finanzas*, 6(7), 67–79.

Marcano Molano, P. G., & Uribe Veintimilla, A. M. (2022). *Satisfacción estudiantil como un indicador de la calidad educativa. Revista Ciencias Pedagógicas e Innovación*, 10(1), 79–84. https://doi.org/10.26423/rcpi.v10i1.520

Martínez, M. M., García, B. & Quintanal, J. (2006). El perfil del profesor universitario de calidad desde la perspectiva del alumnado. *Educación*, 21(9), 183–198.

Mejías, A. & Martínez, D. (2009). Desarrollo de un Instrumento para Medir la Satisfacción Estudiantil en Educación Superior. *Docencia Universitaria*, 10(2), 29–47.

Mejías, A., Reyes, Ó. & Maneiro, N. (2006). Calidad de los servicios en la educación superior mexicana: Aplicación del servqualing en Baja California. *Investigación y Ciencia*, 14(34), 36–41.

Osorio, J. V. & Pérez, K. M. (2010). El nivel de satisfacción escolar y su relación con la orientación vocacional en alumnos de psicología educativa (Tesis profesional, Universidad Pedagógica Nacional, México).

Panera, F. (1999). Calidad total en la enseñanza universitaria. Revista Dirección y Organización, 21, 94–102.

Pérez, R. (2004). La calidad en la educación. En R. Pérez, F. López, M. D. Peralta, & P. M. Fernández. Hacia una educación de calidad. Gestión, instrumentos y evaluación (3 ed.). Madrid: Narcea.

Salinas, A. Morales, J. A. & Martínez, P. (2008). Satisfacción del estudiante y calidad universitaria: un análisis explicatorio en la unidad académica multidisciplinaria agronomía y ciencias de la Universidad Autónoma de Tamaulipas, México. Revista de Enseñanza Universitaria, 23, 39–55. https://institucional.us.es/revistas/universitaria/31/4SalinasGuti.pdf

Sánchez, J. L. (2008). Reformas, investigación, innovación y calidad educativa. *Revista Iberoamericana de Educación*, 46(2), 1–12.

Sandoval, L. Y. (2006). El ser y el hacer de la organización educativa. *Educación y Educadores*, 9(1), 33–53. https://www.redalyc.org/articulo.oa?id=83490104

Sinclaire, J. K. (2014). An empirical investigation of student satisfaction with college courses. *Research in Higher Education Journal*, 23, 1–21.

Surdez-Pérez, E. G., Sandoval-Caraveo, M. C., & Lamoyi-Bocanegra, C. L. (2018). Satisfacción estudiantil en la valoración de la calidad educativa universitaria. *Educación y Educadores*, 21(1), 9–26. https://doi.org/10.5294/edu.2018.21.1.1

UNESCO (2014). Position Paper on Education. Post-2015. https://unesdoc.unesco.org/images/0022/002273/227336E.pdf

UNESCO Institute for Statistics (s. f.). Upper Secondary Education (ISCED level 3) [Glosario]. https://databrowser.uis.unesco.org/resources/glossary/2950

Vázquez, A. (2013). Calidad educativa. *Investigación Educativa*, 17(2).

Vergara, J. C. & Quesada, V. M. (2011). Análisis de la calidad en el servicio y satisfacción de los estudiantes de Ciencias Económicas de la Universidad de Cartagena mediante un modelo de ecuaciones estructurales. *Revista Electrónica de Investigación Educativa*, 13(1), 108–122. https://www.scielo.org.mx/scielo.php?pid=S1607-40412011000100007&script=sci_arttext

7 Food Insecurity and Non-Communicable Diseases

A Bibliometric Review of Diabetes, Obesity, and Hypertension Research

Dalia García-Orozco, Victor Alfaro-Garcia, and Indira C. Ventura-Ruiz

7.1 INTRODUCTION

The global burden of non-communicable diseases (NCDs) has reached unprecedented proportions, accounting for over 74% of all deaths worldwide and representing the greatest challenge to public health systems in the 21st century (World Health Organization's [WHO], 2022). Among these conditions, diet-related NCDs, including type 2 diabetes, obesity, and hypertension, have emerged as the defining health crisis of our era, with prevalence rates that have tripled in the past four decades (WHO, 2023). This epidemiological transition has disproportionately affected low- and middle-income populations, revealing a troubling paradox: those most vulnerable to nutritional deficiencies are simultaneously at greatest risk for conditions once considered diseases of affluence (Phelps et al., 2024).

At the intersection of this paradox lies the phenomenon of food insecurity, the limited or uncertain access to adequate, nutritious food (Food and Agriculture Organization of the United Nations [FAO], 2001). Once conceptualized primarily as a driver of undernutrition, food insecurity has increasingly been recognized as a fundamental social determinant of diet-related chronic diseases (Allen & Feigl, 2017). The mechanisms underlying this relationship are multifaceted: economic constraints that favor energy-dense but nutrient-poor foods, the physiological stress responses triggered by uncertain food availability, disrupted eating patterns, and the psychological burden of resource scarcity. These pathways create a biological and behavioral environment that promotes metabolic dysfunction, weight gain, and cardiometabolic risk accumulation over the life course (Darmon & Drewnowski, 2015).

The scientific interest in this nexus between food insecurity and NCDs has grown exponentially over recent decades, generating a substantial body of evidence across disciplines including epidemiology, nutrition science, public health, economics, and

DOI: 10.1201/9781003724902-8

social medicine (Bhutta et al., 2013; Seligman, Bindman, Vittinghoff, Kanaya, & Kushel, 2007; Seligman, Laraia, & Kushel, 2010). However, the rapid proliferation of research in this domain presents significant challenges for scholars, policymakers, and practitioners seeking to synthesize knowledge, identify research gaps, and understand the intellectual evolution of the field (Donthu, Kumar, Mukherjee, Pandey, & Lim, 2021). Traditional narrative reviews, while valuable, cannot adequately capture the structural and dynamic characteristics of such a rapidly expanding research landscape (Grant & Booth, 2009).

Bibliometric analysis offers a rigorous, quantitative approach to mapping scientific knowledge production (Blanco-Mesa, León-Castro, & Merigó, 2019; García-Orozco, Alfaro-García, Merigó, Espitia Moreno, & Gómez Monge, 2022; Monge, Alfaro-García, Espitia-Moreno, García-Orozco, & de Vivar Mercadillo, 2022). Defined as the statistical analysis of bibliographic data, bibliometrics enables researchers to identify publication patterns, influential works, productive authors and institutions, collaborative networks, and thematic trajectories within a research domain (Donthu, Kumar, Mukherjee, Pandey, & Lim, 2021; García-Orozco, Espitia-Moreno, Alfaro-García, & Merigó, 2020). This methodology, rooted in the field of library and information science, has gained substantial traction across scientific disciplines as a fundamental tool for research evaluation and science mapping. The development of sophisticated software tools particularly the bibliometrix R package and its interactive interface Biblioshiny has democratized access to advanced bibliometric techniques, allowing comprehensive analyses of large-scale datasets with unprecedented precision and reproducibility (Aria & Cuccurullo, 2017; Zhao & Li, 2023).

Despite the growing body of primary research linking food insecurity to diet-related NCDs, no comprehensive bibliometric analysis has yet examined the structure, evolution, and thematic composition of this interdisciplinary field. Such an analysis is essential to identify the foundational works that have shaped the discourse, map the geographic and institutional distribution of knowledge production, detect emerging research frontiers, and reveal collaborative patterns that may facilitate or constrain scientific progress (Newman, 2001; van Eck & Waltman, 2010). Understanding these dimensions is critical not only for researchers seeking to position their work within the broader scholarly conversation but also for funding agencies, academic institutions, and policymakers who require evidence-based assessments of research priorities and knowledge gaps (Callon, Courtial, & Laville, 1991; Small, 1973).

The objective of this study is to provide a comprehensive bibliometric analysis of the scientific literature examining the relationship between food insecurity and diet-related NCDs (diabetes, obesity, and hypertension) published between 1991 and 2025. Specifically, this analysis aims to: (1) characterize the temporal evolution and growth dynamics of research output in this domain; (2) identify the most productive and influential journals, authors, institutions, and countries contributing to the field; (3) map the intellectual structure through co-occurrence network analysis of author keywords; (4) delineate thematic clusters and their strategic positioning within the research landscape; and (5) trace the citation pathways that connect foundational works to contemporary research through historiographic analysis. Through these objectives, this study seeks to provide both a retrospective synthesis and a forward-looking assessment that can inform future research directions in this critical area of public health inquiry.

7.2 METHODS

This bibliometric analysis was conducted following the Preferred Reporting Items for Systematic Reviews and Meta-Analyses (PRISMA) 2020 guidelines adapted for bibliometric studies (Yepes-Nuñez, Urrútia, Romero-García, & Alonso-Fernández, 2021). Scopus was selected as the primary database due to its comprehensive coverage of peer-reviewed literature in health sciences and nutrition, standardized metadata structure, and suitability for large-scale bibliometric analyses (Mongeon & Paul-Hus, 2016). Figure 7.1 shows the process that was followed.

A systematic search was performed on January 16, 2026, employing the following query: TITLE-ABS-KEY (("food insecurity" OR "food security") AND (diabetes OR "type 2 diabetes" OR obesity OR overweight OR hypertension OR "high blood pressure")) AND PUBYEAR > 1990 AND PUBYEAR < 2026 AND (LIMIT-TO (DOCTYPE, "ar") OR LIMIT-TO(DOCTYPE, "re")) AND (EXCLUDE (PUBSTAGE, "aip")). The search strategy was designed to capture the intersection between food insecurity and diet-related NCDs, restricting results to original research articles and reviews published between 1991 and 2025, while excluding articles in press to ensure complete bibliographic indexing. No language restrictions were applied.

The initial search yielded 3,650 records exported in Comma-Separated Values (CSV) format. Duplicate detection was performed using both Digital Object Identifier (DOI) and Scopus unique identifier (electronic Identifier, EID); of the total records, 3,507 (96.1%) contained DOIs with two potential duplicates identified, while EID analysis

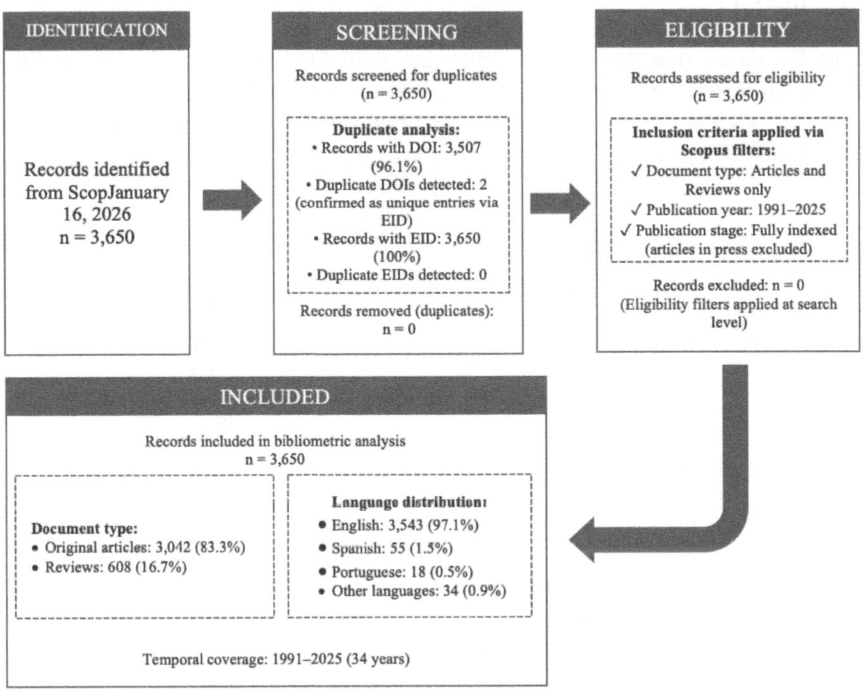

FIGURE 7.1 PRISMA 2020 flow diagram for bibliometric analysis.

confirmed all 3,650 entries as unique records in the database. Consequently, no records were removed during the deduplication process, resulting in a final analytical corpus of 3,650 documents comprising 3,042 original articles (83.3%) and 608 reviews (16.7%).

Bibliometric analyses were performed using the bibliometrix R package within R statistical software, an open-source tool providing comprehensive algorithms for science mapping and quantitative bibliographic analysis (Aria & Cuccurullo, 2017). Descriptive metrics included temporal publication trends, document type and language distributions, and geographic patterns. Productivity indicators were calculated for journals, authors, institutions, and countries, encompassing total publications (TP), total citations (TC), citations per publication (C/P), and h-index values.

Co-occurrence analysis was conducted on author keywords to construct a network of thematic relationships, applying the Louvain algorithm for community detection with visualization of the 50 most frequent terms and a minimum edge weight of three co-occurrences (Blondel, Guillaume, Lambiotte, & Lefebvre, 2008). A thematic map (strategic diagram) was generated through co-word analysis with a minimum frequency threshold of 10 occurrences, classifying research themes according to their centrality (relevance within the field) and density (internal cohesion). Finally, a historiograph depicting direct citation relationships among the 30 most influential publications was constructed to trace the intellectual evolution of the research domain. All visualizations were rendered using the Fruchterman-Reingold layout algorithm and exported at 120 dpi resolution (Fruchterman & Reingold, 1991).

7.3 RESULTS

Table 7.1 shows that this field of study has expanded rapidly and is increasingly consolidating. Covering 1991–2025, the corpus comprises 3,650 documents disseminated across 1,178 sources, indicating a highly diversified outlet structure rather than concentration in a small set of core journals. This dispersion is consistent with an

TABLE 7.1
Main Characteristics of the Analyzed Dataset

Timespan	**1991:2025**
Sources (journals, books, etc.)	1,178
Documents	3,650
Annual growth rate %	20.04
Document average age	6.06
Average citations per doc	28.91
References	24,294
Authors	12,605
Single-authored docs	0
Co-authors per doc	11.3
International co-authorships %	21.7
Article	3,042
Review	608

inherently interdisciplinary topic spanning nutrition, epidemiology, public health, health policy, and social determinants research.

The estimated annual growth rate of 20.04% signals strong acceleration, suggesting that the linkage between food insecurity or food security and cardiometabolic conditions has become a high-priority agenda over the last decade. The average document age of 6.06 years reinforces the recency of the evidence base and implies that conceptual and methodological approaches are evolving quickly. Despite this youth, the mean of 28.91 citations per document indicates substantial scholarly uptake, suggesting that the literature is not merely expanding in volume but is also being actively integrated into subsequent studies and syntheses.

The dataset includes 24,294 cited references, showing an interesting cumulative knowledge foundation that supports hypothesis building around behavioral, nutritional, and structural pathways connecting food insecurity with diabetes, obesity, overweight, and hypertension. Collaboration is a predominant feature with 12,605 authors contributing to the corpus, and the average of 11.3 co-authors per document points to team-based production typical of large-scale surveys, cohort studies, and multi-institutional projects. International co-authorship stands at 21.7%, evidencing meaningful cross-border cooperation while also highlighting unrealized potential for more globally comparative work.

Graphic 7.1 shows the production and average total citations per year. Results show that the scientific production is modest through the 1990s and early 2000s, then begins a sustained climb that becomes steep after the mid-2010s and accelerates sharply from 2019 onward, reaching the highest levels by 2025. This pattern is consistent with the topic's transition from an emergent linkage into a mainstream research stream that is now producing at scale.

The mean citation shows prominent spikes in the late 1990s and mid-2000s, followed by a more moderate development and then a decline toward 2025. This is a classic bibliometric "age and denominator" effect: older, foundational papers have had more time to accumulate citations and can concentrate attention, producing

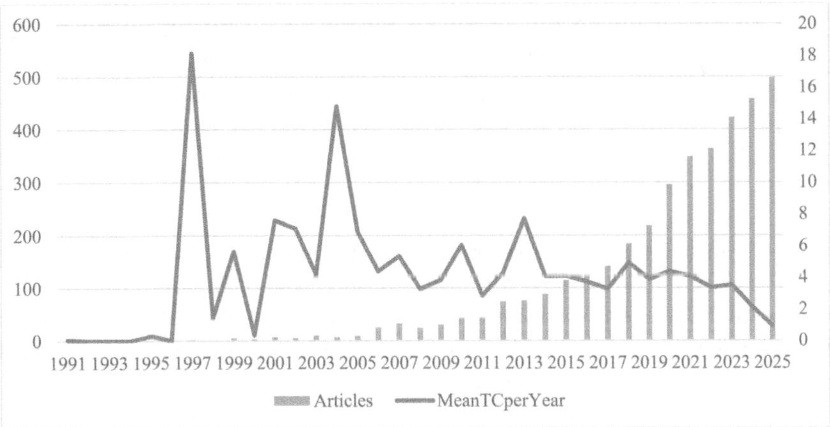

GRAPHIC 7.1 Annual scientific production on food insecurity and non-communicable diseases (1991–2025).

sharp peaks when the annual volume of publications is still low. As the field expands rapidly, citations disperse across many more papers and recent publications have had insufficient time to accrue citations (Bornmann & Leydesdorff, 2014; Price, 1965).

Taken together, the graphic suggests a two-phase evolution. An early phase dominated by a small number of highly influential contributions that shaped measurement and conceptual framing, and a second phase characterized by rapid scale-up, diversification of methods, and broader thematic spread across diabetes, obesity, and hypertension pathways. The rising production with falling average citation also signals a potential pivot in the field: from establishing associations and definitions toward building a larger, policy-facing evidence base, where influence may be better captured by synthesis, guidelines, and translational uptake rather than immediate citation velocity.

Table 7.2 offers a journal-level map of where the food insecurity–NCD literature is being published and, more importantly, how influence is distributed across outlets when productivity is evaluated alongside citation-based performance indicators. Performance analysis in bibliometrics is typically reported through a combination of documents, citations, h-index, cites per paper, and citation thresholds because each captures a different dimension of visibility and scholarly uptake (Donthu, Kumar, Mukherjee, Pandey, & Lim, 2021; Hirsch, 2005).

Three outlet roles emerge clearly. First, high-volume platforms such as *Nutrients* (131 papers), *International Journal of Environmental Research and Public Health* (93), *PLOS One* (75), and *BMC Public Health* (47) function as the field's main dissemination infrastructure. Their contribution is breadth: they accommodate heterogeneous designs, settings, and populations, enabling rapid accumulation of empirical evidence across nutrition, public health, and social determinants. Importantly, their influence is not driven by single "blockbuster" articles; rather, a substantial share of papers reaches the baseline citation threshold, indicating sustained, cumulative relevance across many contributions.

Second, a smaller set of clinically anchored journals concentrates high-intensity impact despite lower output. *Pediatrics* and *Journal of Nutrition* show very high citations per paper and large counts of papers exceeding 100 citations, signaling that highly reused contributions often sit at the intersection of food insecurity with developmental risk, diet quality, and long-run cardiometabolic trajectories. *Diabetes Care* and *Obesity* show a similar pattern: repeated high-citation uptake across multiple papers indicates that food insecurity is being treated as clinically consequential within diabetes and obesity research agendas, not as a peripheral socioeconomic covariate.

Third, translational bridges connect etiologic evidence to prevention and policy. Outlets such as *Public Health Nutrition*, *Appetite*, *American Journal of Preventive Medicine*, and *Preventing Chronic Disease* combine steady throughput with meaningful citation depth, consistent with work focused on behavioral pathways, intervention design, and scalable public health responses.

Table 7.3 identifies the citation "anchor texts" that have shaped this corpus, and it shows a field built on three tightly connected knowledge streams rather than a single, food-insecurity-only tradition. Interpreting highly cited documents is useful precisely because citations operate as signals of perceived importance in the cumulative conversation of a domain (Garfield, 1972).

TABLE 7.2
Most Productive Journals in Food Insecurity and NCDs Research

R	Journal Name	TP	TC	C/P	H	≤100	≤10
1	*Nutrients*	131	2,489	19	24	5	57
2	*International Journal of Environmental Research and Public Health*	93	1,406	15.12	20	0	38
3	*PLOS One*	75	857	11.43	16	1	25
4	*Public Health Nutrition*	72	2,507	34.82	26	3	50
5	*Journal of Hunger and Environmental Nutrition*	65	893	13.74	16	1	29
6	*Appetite*	58	1,751	30.19	25	4	42
7	*Journal of Nutrition*	53	7,973	150.43	36	20	71
8	*BMC Public Health*	47	749	15.94	16	1	24
9	*BMJ Open*	39	580	14.87	13	0	16
10	*Frontiers in Public Health*	32	494	15.44	8	1	6
11	*Preventive Medicine Reports*	31	463	14.94	12	1	12
12	*Frontiers in Nutrition*	29	449	15.48	13	0	13
13	*Journal of the Academy of Nutrition and Dietetics*	29	1,454	50.14	17	6	22
14	*Journal of General Internal Medicine*	28	1,371	48.96	11	4	12
15	*Journal of Nutrition Education and Behavior*	27	0	0	0	0	0
16	*Behavioral and Brain Sciences*	26	297	11.42	5	1	1
17	*Diabetes Care*	26	3,486	134.08	19	10	20
18	*Pediatrics*	24	5,022	209.25	31	17	45
19	*Ciencia E Saude Coletiva*	22	310	14.09	10	0	10
20	*Contemporary Clinical Trials*	22	186	8.45	7	0	4
21	*Current Developments in Nutrition*	22	602	27.36	10	1	10
22	*JAMA Network Open*	22	382	17.36	11	1	12
23	*Journal of the American Dietetic Association*	22	2,184	99.27	19	7	20
24	*Obesity*	22	2,863	130.14	29	6	67
25	*Preventing Chronic Disease*	22	685	31.14	14	1	16
26	*American Journal of Preventive Medicine*	21	713	33.95	12	1	12
27	*Social Science and Medicine*	21	521	24.81	11	1	11
28	*American Journal of Clinical Nutrition*	19	3,084	162.32	13	5	14
29	*Current Nutrition Reports*	19	1,295	68.16	12	6	13
30	*Academic Pediatrics*	18	0	0	0	0	0
31	*Childhood Obesity*	18	0	0	0	0	0
32	*Pediatric Obesity*	18	0	0	0	0	0
33	*Asia Pacific Journal of Clinical Nutrition*	17	625	36.76	13	1	13
34	*Food Security*	17	1,425	83.82	15	3	18
35	*British Journal of Nutrition*	16	0	0	0	0	0
36	*Canadian Journal of Diabetes*	16	206	12.88	8	0	7
37	*BMC Nutrition*	15	103	6.87	6	0	3
38	*Translational Behavioral Medicine*	15	334	22.27	10	1	10
39	*Food and Nutrition Bulletin*	14	580	41.43	11	2	11
40	*Journal of Immigrant and Minority Health*	14	599	42.79	11	1	11

Note: R: Ranking; TP: Total papers; TC: Total cites; C/P: Total cites divided by total papers; H: H-index; ≤100: number of articles with more than 100 citations; ≤10: number of articles with more than 10 citations

TABLE 7.3
Most Cited Articles in Food Insecurity and NCDs Research

R	TC	J	Title	Author/s	Year	C/Y
1	2,554	*Science*	Climate change impacts on global food security	Wheeler, T.; von Braun, J.	2013	196.5
2	2,159	*Lancet*	Evidence-based interventions for improvement of maternal and child nutrition: What can be done and at what cost?	Bhutta, Z.A.; Das, J.K.; Rizvi, A.; Gaffey, M.F.; Walker, N.; Horton, S.; Webb, P.; Lartey, A.; Black, R.E.	2013	166.1
3	1,990	*Am. J. Clin. Nutr.*	Poverty and obesity: The role of energy density and energy costs	Drewnowski, A.; Specter, S.E.	2004	90.5
4	1,279	*J. Nutr.*	Food insecurity is associated with chronic disease among low-income NHANES participants	Seligman, H.K.; Laraia, B.; Kushel, M.B.	2010	79.9
5	1,234	*Diabetes Care*	Social determinants of health and diabetes: A scientific review	Hill-Briggs, F.; Adler, N.E.; Berkowitz, S.A.; Chin, M.H.; Gary-Webb, T.L.; Navas-Acién, A.; Thornton, P.L.; Haire-Joshu, D.	(2020)	246.8
6	1,036	*Nutr. Rev.*	The nutrition transition: New trends in the global diet	Drewnowski, A.; Popkin, B.M.	1997	35.7
7	992	*Pediatrics*	Clinical Practice Guideline for the Evaluation and Treatment of Children and Adolescents with Obesity	Hampl, S.E.; Hassink, S.G.; Skinner, A.C.; Armstrong, S.C.; Barlow, S.E.; Bolling, C.F.; Edwards, K.C.A.; Eneli, I.; Hamre, R.; Joseph, M.M.; Lunsford, D.; Mendonça, E.; Michalsky, M.P.; Mirza, N.; Ochoa, E.R.; Sharifi, M.; Staiano, A.E.; Weedn, A.E.; Flinn, S.K.; Lindros, J.; Okechukwu, K.	2023	330.7
8	945	*Int. J. Food Microbiol.*	Food-borne diseases - The challenges of 20 years ago still persist while new ones continue to emerge	Newell, D.G.; Koopmans, M.; Verhoef, L.; Duizer, E.; Aïdara-Kane, A.; Sprong, H.; Opsteegh, M.; Langelaar, M.; Threfall, J.; Scheutz, F.; der Giessen, J.V.; Kruse, H.	2010	59.1
9	739	*Annu. Rev. Plant Biol.*	Trends in Global Agricultural Land Use: Implications for Environmental Health and Food Security	Ramankutty, N.; Mehrabi, Z.; Waha, K.; Jarvis, L.; Kremen, C.; Herrero, M.; Rieseberg, L.H.	2018	92.4

(Continued)

TABLE 7.3 ***(Continued)***
Most Cited Articles in Food Insecurity and NCDs Research

R	TC	J	Title	Author/s	Year	C/Y
10	730	*Lancet Planet. Health*	Health and nutritional aspects of sustainable diet strategies and their association with environmental impacts: a global modelling analysis with country-level detail	Springmann, M.; Wiebe, K.; Mason-D'Croz, D.; Sulser, T.B.; Rayner, M.; Scarborough, P.	2018	91.2
11	696	*J. Nutr.*	Food insecurity affects school children's academic performance, weight gain, and social skills	Jyoti, D.F.; Frongillo, E.A.; Jones, S.J.	2005	33.1
12	653	*J. Gen. Intern. Med.*	Food insecurity is associated with diabetes mellitus: Results from the National Health Examination and Nutrition Examination Survey (NHANES) 1999–2002	Seligman, H.K.; Bindman, A.B.; Vittinghoff, E.; Kanaya, A.M.; Kushel, M.B.	2007	34.4
13	646	*Lancet*	The double burden of malnutrition: aetiological pathways and consequences for health	Wells, J.C.; Sawaya, A.L.; Wibaek, R.; Mwangome, M.; Poullas, M.S.; Yajnik, C.S.; Demaio, A.	2020	107.7
14	593	*World Dev.*	Urbanization, lifestyle changes and the nutrition transition	Popkin, B.M.	1999	22
15	585	*J. Nutr.*	Food insecurity is positively related to overweight in women	Townsend, M.S.; Peerson, J.; Love, B.; Achterberg, C.; Murphy, S.P.	2001	23.4
16	575	*J. Nutr.*	Household food insufficiency is associated with poorer health	Vozoris, N.T.; Tarasuk, V.	2003	25
17	493	*J. Am. Diet.* Assoc.	The Food Insecurity-Obesity Paradox: A Review of the Literature and the Role Food Stamps May Play	Dinour, L.M.; Bergen, D.; Yeh, M.-C.	2007	25.9
18	470	*Br. J. Nutr.*	Sex differences in fat storage, fat metabolism, and the health risks from obesity: Possible evolutionary origins	Power, M.L.; Schulkin, J.	2008	26.1

(Continued)

TABLE 7.3 ***(Continued)***
Most Cited Articles in Food Insecurity and NCDs Research

R	TC	J	Title	Author/s	Year	C/Y
19	462	*Lancet*	The political origins of health inequity: Prospects for change	Ottersen, O.P.; Dasgupta, J.; Blouin, C.; Buss, P.; Chongsuvivatwong, V.; Frenk, J.; Fukuda-Parr, S.; Gawanas, B.P.; Giacaman, R.; Gyapong, J.; Leaning, J.; Marmot, M.; McNeill, D.; Mongella, G.I.; Moyo, N.; Mogedal, S.; Ntsaluba, A.; Ooms, G.; Bjertness, E.; Lie, A.L.; Moon, S.; Roalkvam, S.; Sandberg, K.I.; Scheel, I.B.	2014	38.5
20	451	*Global Food Secur.*	Concepts and critical perspectives for food environment research: A global framework with implications for action in low- and middle-income countries	Turner, C.; Aggarwal, A.; Walls, H.; Herforth, A.; Drewnowski, A.; Coates, J.; Kalamatianou, S.; Kadiyala, S.	2018	56.4
21	425	*Public Health Nutr.*	An accelerated nutrition transition in Iran	Ghassemi, H.; Harrison, G.; Mohammad, K.	2002	17.7
22	410	*J. Am. Coll. Cardiol.*	Food Consumption and its Impact on Cardiovascular Disease: Importance of Solutions Focused on the Globalized Food System A Report from the Workshop Convened by the World Heart Federation	Anand, S.S.; Hawkes, C.; de Souza, R.J.; Mente, A.; Dehghan, M.; Nugent, R.; Zulyniak, M.A.; Weis, T.; Bernstein, A.M.; Krauss, R.M.; Kromhout, D.; Jenkins, D.J.A.; Malik, V.; Martínez-González, M.A.; Mozaffarian, D.; Yusuf, S.; Willett, W.C.; Popkin, B.M.	2015	37.3
23	393	*PLoS Med.*	Big food, food systems, and global health	Stuckler, D.; Nestle, M.	2012	28.1
24	383	*J. Community Health*	Exploring mediators of food insecurity and obesity: A review of recent literature	Franklin, B.; Jones, A.; Love, D.; Puckett, S.; Macklin, J.; White-Means, S.	2012	27.4
25	371	*Curr. Biol.*	An evolutionary perspective on food and human taste	Breslin, P.A.S.	2013	28.5

Note: J: Journal; C/Y: Total cites in food insecurity and NCDs divided by years.

First, the top of the list is dominated by macro-level food security scholarship that frames downstream NCD risk through systems pressure. The most cited article, on climate change and global food security (Wheeler & Von Braun, 2013), and the highly cited Lancet evidence synthesis on maternal and child nutrition interventions (Bhutta et al., 2013) position food (in)security as a structural constraint that shapes diet quality, growth, and long-run metabolic vulnerability. This matters for your review because it clarifies why the food insecurity–NCD link is increasingly discussed alongside climate resilience, food systems, and policy feasibility, not only household deprivation.

Second, an obesity-centered conceptual backbone is evident. Drewnowski's work on poverty and obesity (Drewnowski & Specter, 2004) and the "nutrition transition" lineage (Drewnowski & Popkin, 1997; Popkin, 1999, 2002) explain how economic constraints, energy density, and rapid dietary change create predictable pathways toward overweight and cardiometabolic risk. The persistence of the food insecurity–obesity paradox theme shows that the field's enduring puzzle is not whether scarcity and obesity can co-exist, but under which structural and behavioral conditions they co-produce risk (Dietz, 1995; Dinour, Bergen, & Yeh, 2007; Townsend, Peerson, Love, Achterberg, & Murphy, 2001).

Third, the NCD connection is operationalized through high-visibility epidemiologic evidence and clinical translation. Multiple National Health and Nutrition Examination Survey (NHANES)-based papers linking food insecurity to chronic disease and diabetes (Seligman, Bindman, Vittinghoff, Kanaya, & Kushel, 2007, 2010) serve as methodological reference points because they offer scalable measurement and generalizable inference. The more recent "social determinants and diabetes" scientific review (Hill-Briggs et al., 2020) and pediatric obesity guideline (Hampl et al., 2023) exhibit extremely high annual citation rates, indicating rapid uptake and institutionalization into clinical and policy discourse.

Two implications follow. Hypertension is comparatively less visible among the top-cited set, suggesting an opportunity for synthesis work that treats blood pressure outcomes with the same conceptual depth achieved for obesity and diabetes. Also, citation profiles are time-sensitive; recent guideline-type documents can accumulate citations quickly, while newer empirical papers may appear undercited simply due to shorter exposure time (Burrell, 2002; Phelps et al., 2024).

Table 7.4 profiles the authors who structure this literature through a combination of volume, citation impact, and sustained influence. Interpreting these indicators jointly is standard in performance analysis because publication counts capture productivity, whereas citations and the h-index better reflect depth and durability of scholarly uptake.

Two features stand out. First, the highest-ranked authors combine repeated output with concentrated impact around the clinical and policy interface of food insecurity. Seligman (26 papers; 4,802 citations; h-index 25) and Berkowitz (33 papers; 3,785 citations; h-index 20) anchor a core stream linking food insecurity with diabetes-related outcomes, screening, and care delivery (Hill-Briggs et al., 2020; Seligman, Bindman, Vittinghoff, Kanaya, & Kushel, 2007, 2010). Their citation-per-paper levels are high and, critically, they show many papers crossing the "high-impact" threshold (13 papers with at least 100 citations for each), suggesting influence that is distributed across multiple contributions rather than driven by a single landmark item. This pattern matches the rationale for author rankings that integrate productivity with citation-based influence to identify both prolific and field-shaping scholars.

TABLE 7.4
Most Productive and Influential Authors in the Field

R	Author Name	Affiliation	TP	TC	H	C/P	≤100	≤10
1	Berkowitz, S.A.	Harvard Medical School, Boston, MA, United States	33	3,785	20	114.7	13	25
2	Frongillo, E.A.	Department of Preventive Medicine, Hanyang University, Seoul, South Korea	26	2,077	15	79.88	5	19
3	Seligman, H.K.	Department of Agricultural and Consumer Economics, University of Illinois Urbana-Champaign, Urbana, IL, United States	26	4,802	25	184.69	13	27
4	Egede, L.E.	Department of Medicine, Medical College of Wisconsin, Milwaukee, WI, United States	25	797	13	31.88	1	15
5	Liese, A.D.	University of Washington School of Medicine, Seattle, WA, United States	24	678	10	28.25	1	10
6	Pérez-Escamilla, R.	Instituto de Nutrição Josué de Castro, Universidade Federal do Rio de Janeiro, Rio de Janeiro, RJ, Brazil	20	1,390	16	69.5	4	17
7	Walker, R.J.	Department of Society, Human Development and Health, Harvard T.H. Chan School of Public Health, Boston, MA, United States	19	3,224	17	169.68	3	23
8	Peltzer, K.	Department of Psychology, University of the Free State, Bloemfontein, Free State, South Africa	17	118	5	6.94	0	3
9	Leung, C.W.	New America Foundation, Washington, D.C., DC, United States	16	729	13	45.56	2	15
10	Mozaffarian, D.	Tufts University, Medford, MA, United States	16	749	8	46.81	2	7
11	Black, M.M.	Department of Organizational Systems and Adult Health, University of Maryland School of Nursing, Baltimore, MD, United States	15	3,330	14	222	5	14
12	Mendoza, J.A.	University of Washington School of Medicine, Seattle, WA, United States	15	161	6	10.73	0	4
13	Pengpid, S.	Department of Psychology, College of Medical and Health Science, Taichung, Taiwan	15	118	5	7.87	0	3
14	Laraia, B.A.	Nutrition Policy Institute, University of California Division of Agriculture and Natural Resources, Davis, CA, United States	14	2,033	11	145.21	4	12
15	Rosas, L.G.	Department of Epidemiology and Population Health, Stanford University School of Medicine, Stanford, CA, United States	14	452	11	32.29	1	11

(Continued)

TABLE 7.4 ***(Continued)***
Most Productive and Influential Authors in the Field

R	Author Name	Affiliation	TP	TC	H	C/P	≤100	≤10
16	Pihoker, C.	Department of Pediatrics, University of Washington School of Medicine, Seattle, WA, United States	13	159	6	12.23	0	4
17	Reboussin, B.A.	South College, Atlanta, GA, United States	13	92	5	7.08	0	3
18	Gundersen, C.	Agricultural Marketing in the Department of Agricultural and Consumer Economics, University of Illinois Urbana-Champaign, Urbana, IL, United States	12	1,368	13	114	7	13
19	Gross, R.S.	Division of General Pediatrics, NYU Grossman School of Medicine, New York, NY, United States	11	466	10	42.36	2	10
20	McElfish, P.A.	Center for Nutrition and Health Impact, Omaha, NE, United States	11	85	5	7.73	0	3
21	Ritchie, L.D.	University of Colorado Anschutz School of Medicine, Aurora, CO, United States	11	157	6	14.27	0	5
22	Sauder, K.A.	South College, Atlanta, GA, United States	11	73	4	6.64	0	2
23	Taveras, E.M.	Department of Pediatrics, Massachusetts General Hospital, Boston, MA, United States	11	88	5	8	0	3
24	Basu, S.	Division of General Medicine and Clinical Epidemiology, UNC School of Medicine, Chapel Hill, NC, United States	10	577	7	57.7	2	7
25	Crews, D.C.	Johns Hopkins School of Nursing, Baltimore, MD, United States	10	355	7	35.5	2	6
26	Long, C.R.	Programa de Pós-Graduação em Ciências da Saúde, Universidade de Caxias do Sul, Caxias do Sul, RS, Brazil	10	259	8	25.9	0	7
27	Messito, M.J.	Division of General Pediatrics, NYU Grossman School of Medicine, New York, NY, United States	10	251	6	25.1	1	5
28	Myers, C.A.	Division of General Internal Medicine, UCSF School of Medicine, San Francisco, CA, United States	10	547	9	54.7	2	9
29	Shamah-Levy, T.	Yale University, New Haven, CT, United States	10	352	8	35.2	0	8
30	Bateson, M.	Biosciences Institute, Newcastle University, Newcastle, Tyne and Wear, United Kingdom	9	328	5	36.44	1	4

Second, Table 7.4 reveals a strong pediatric and life-course substream, where fewer papers can still yield substantial influence. Black (15 papers; 3,330 citations; the highest citations per paper in this list) and Frongillo (26 papers; 2,077 citations) reflect an enduring emphasis on maternal-child nutrition, developmental vulnerability, and the downstream metabolic consequences of constrained food access (Bhutta et al., 2013; Jyoti, Frongillo, & Jones, 2005). The prominence of these authors, alongside Liese, Mendoza, Pihoker, Gross, Messito, and Taveras, indicates that pediatric settings are not peripheral to the NCD framing; they are central entry points where food insecurity is operationalized as an upstream risk factor with long-run consequences.

Institutionally, the leading affiliations cluster in major U.S. academic centers (Harvard, Illinois Urbana-Champaign, Stanford, Johns Hopkins, Tufts, University of Washington, NYU), indicating a concentration of agenda-setting capacity in U.S.-based research networks. At the same time, the presence of Pérez-Escamilla and Long (Brazil), Peltzer (South Africa), Pengpid (Taiwan), and Bateson (United Kingdom) signals that the field also includes globally distributed contributors, although with more heterogeneous citation profiles.

Finally, contrasts within the table are informative. Authors such as Peltzer and Pengpid publish repeatedly but with low citation density and limited high-citation papers, consistent with contributions that expand geographic coverage or survey breadth without becoming central reference points in the core food insecurity–NCD discourse. Overall, the table depicts a mature field in which a relatively small set of authors provides the conceptual and empirical backbone, while a wider perimeter of contributors broadens populations, settings, and external validity.

Table 7.5 shows a highly concentrated institutional backbone for food insecurity and NCDs research, with agenda-setting capacity clustered in a small number of U.S. academic medical centers and schools of public health. This pattern is typical

TABLE 7.5
Leading Institutions in Food Insecurity and NCDs Research

R	Institution	Country	TP	TC	H	C/P	≤100	≤10
1	University of California, San Francisco	United States	114	5,320	29	46.67	14	40
2	UCSF School of Medicine	United States	88	3,755	21	42.67	9	40
3	Harvard T.H. Chan School of Public Health	United States	84	3,066	24	36.5	8	39
4	Harvard Medical School	United States	80	3,212	25	40.15	13	39
5	The University of North Carolina at Chapel Hill	United States	76	1,107	10	14.57	2	11
6	University of South Carolina	United States	68	2,774	25	40.79	4	41
7	University of Michigan, Ann Arbor	United States	64	2,378	18	37.16	4	21
8	University of Minnesota Twin Cities	United States	59	736	16	12.47	0	20
9	Massachusetts General Hospital	United States	58	2,428	22	41.86	9	30

(Continued)

TABLE 7.5 *(Continued)*
Leading Institutions in Food Insecurity and NCDs Research

R	Institution	Country	TP	TC	H	C/P	≤100	≤10
10	Johns Hopkins Bloomberg School of Public Health	United States	55	3,048	16	55.42	2	25
11	University of Washington	United States	52	4,105	21	78.94	7	31
12	Tufts University	United States	51	5,228	21	102.51	13	31
13	Johns Hopkins University	United States	49	2,798	19	57.1	6	28
14	UNC School of Medicine	United States	48	2,254	15	46.96	4	19
15	UNC Gillings School of Global Public Health	United States	47	705	11	15	3	12
16	Yale University	United States	46	2,042	23	44.39	7	27
17	University of Toronto	Canada	45	1,770	20	39.33	4	27
18	Johns Hopkins University School of Medicine	United States	44	2,568	16	58.36	6	23
19	London School of Hygiene & Tropical Medicine	United Kingdom	44	2,182	18	49.59	6	24
20	University of Pennsylvania	United States	43	1,059	17	24.63	2	22
21	Wake Forest University School of Medicine	United States	42	591	15	14.07	0	18
22	University of Michigan School of Public Health	United States	41	790	15	19.27	2	18
23	Pennsylvania State University	United States	40	1,612	16	40.3	6	19
24	University of the Witwatersrand, Johannesburg	South Africa	40	759	12	18.98	1	16
25	Stanford University School of Medicine	United States	40	882	14	22.05	2	16
26	National Institutes of Health NIH	United States	39	0	0	0	0	0
27	Northwestern University Feinberg School of Medicine	United States	39	536	12	13.74	0	16
28	University of Cape Town	South Africa	38	773	16	20.34	1	19
29	University of Pennsylvania Perelman School of Medicine	United States	38	777	14	20.45	2	15
30	South African Medical Research Council	South Africa	38	598	12	15.74	2	12

of mature, policy-facing health domains where large clinical datasets, national surveys, and implementation platforms are housed in research-intensive universities. In bibliometric performance analysis, institutional standing is more credibly inferred from the combined reading of productivity, total citations, and the h-index, because high output alone can reflect scale, whereas citation depth indicates sustained uptake across multiple papers (Merigó & Yang, 2017).

At the top of the ranking, the University of California system is structurally dominant. "University of California, San Francisco" leads in volume and cumulative influence (114 papers; 5,320 citations; h-index 29), while "UCSF School of Medicine" also ranks second (88 papers; 3,755 citations; h-index 21). The co-presence of these two entries is analytically meaningful: it suggests that the field's core production is not just university-wide but heavily embedded in medical-school ecosystems, where food insecurity is operationalized as a clinical risk factor linked to diabetes, obesity, and cardiometabolic outcomes. It also signals a technical point for the methods section: affiliation harmonization matters, because institutional variants can split performance across labels, and this can alter rank ordering if not normalized by affiliation ID.

Harvard-affiliated institutions occupy another central pole. Harvard T.H. Chan School of Public Health (84 papers; 3,066 citations; h-index 24) and Harvard Medical School (80 papers; 3,212 citations; h-index 25) combine high productivity with strong citation depth, consistent with a research profile that spans epidemiology, health equity, and clinical translation. The presence of Massachusetts General Hospital (58 papers; 2,428 citations; h-index 22) reinforces the medical-center role in moving food insecurity from social exposure to actionable screening and intervention workflows within care delivery.

Impact concentration varies across institutions in ways that reveal how influence is produced. Tufts University and the University of Washington have lower output than UCSF but markedly higher citation density (Tufts: 51 papers; 5,228 citations; University of Washington: 52 papers; 4,105 citations), implying that these sites contribute a smaller number of disproportionately field-shaping papers, often characteristic of guideline-relevant syntheses or highly reusable empirical designs.

Finally, the institutional geography is not exclusively U.S.-centric. The University of Toronto and the London School of Hygiene and Tropical Medicine show strong positions, while South African institutions (University of the Witwatersrand, University of Cape Town, South African Medical Research Council) indicate an active global-health strand where food insecurity intersects with NCD risk under different structural constraints. The NIH entry with zero citations should be treated as a data-quality signal rather than a substantive conclusion, reinforcing the need to validate citation capture and affiliation parsing before drawing institutional performance claims.

Table 7.6 characterizes the geographic structure of the field by combining national output, citation impact, and population-adjusted indicators. The distribution is strongly asymmetric. The United States dominates in absolute terms, producing 2,062 publications and 67,977 citations, with the highest h-index in the table. This is not merely a scale effect: the United States also shows a high productivity rate per capita (about 6.14 publications per million inhabitants) and a large citation density per capita (about 202.43 citations per million), indicating both volume capacity and sustained uptake.

A second tier is formed by the United Kingdom, Canada, and Australia, which publish far fewer papers than the United States but exhibit comparable or higher average influence per article. The United Kingdom (300 papers) and Italy (76 papers) show particularly strong citations per paper, suggesting that their contributions concentrate in highly cited syntheses, cohort evidence, or policy-relevant frameworks that diffuse widely across the literature. Canada combines a solid h-index with the highest publications-per-capita among large countries listed (about 6.33 per million), signaling efficient production relative to population size.

TABLE 7.6
Countries' Scientific Production on Food Insecurity and NCDs

R	Country	TP	TC	H	C/P	≤100	≤10	Population	P/Po	C/Po
1	United States	2,062	67,977	116	32.97	153	1,022	335.8	6.14	202.43
2	United Kingdom	300	14,698	49	48.99	28	183	67.7	4.43	217.1
3	Canada	247	10,335	42	41.84	16	142	39	6.33	265
4	South Africa	171	3,847	33	22.5	5	80	60	2.85	64.12
5	Australia	167	7,402	42	44.32	17	98	26.5	6.3	279.32
6	Brazil	164	3,537	27	21.57	4	64	212.6	0.77	16.64
7	India	106	4,482	27	42.28	7	64	1,441	0.07	3.11
8	Iran	98	1,560	19	15.92	2	41	89	1.1	17.53
9	Mexico	87	1,897	22	21.8	3	43	128	0.68	14.82
10	China	79	1,968	21	24.91	6	30	1,425	0.06	1.38
11	Italy	76	3,826	29	50.34	11	49	59	1.29	64.85
12	Ethiopia	61	946	17	15.51	2	25	126.5	0.48	7.48
13	Germany	57	4,029	18	70.68	4	23	83.2	0.69	48.43
14	France	55	1,954	24	35.53	4	34	68	0.81	28.74
15	South Korea	52	791	14	15.21	1	23	51.7	1.01	15.3
16	Spain	50	1,065	18	21.3	1	20	48.5	1.03	21.96
17	Colombia	45	598	12	13.29	1	14	52	0.87	11.5
18	Malaysia	45	1,101	15	24.47	3	16	33	1.36	33.36
19	Netherlands	45	2,849	18	63.31	7	23	17.5	2.57	162.8
20	Kenya	43	1,374	16	31.95	2	22	54	0.8	25.44
21	Nigeria	42	695	16	16.55	0	24	223	0.19	3.12
22	Switzerland	40	2,493	17	62.33	7	23	8.8	4.55	283.3
23	New Zealand	39	849	16	21.77	0	24	5.1	7.65	166.47
24	Thailand	38	1,047	13	27.55	2	13	70	0.54	14.96
25	Ghana	36	3,159	15	87.75	2	17	34	1.06	92.91
26	Belgium	34	1,321	15	38.85	5	20	11.6	2.93	113.88
27	Sweden	33	1,041	16	31.55	2	25	10.5	3.14	99.14
28	Pakistan	32	3,043	14	95.09	3	17	240	0.13	12.68
29	Bangladesh	31	308	9	9.94	0	8	172	0.18	1.79
30	Denmark	29	2,428	16	83.72	4	20	5.9	4.92	411.53

Population normalization highlights small, high-income countries as disproportionate contributors. Denmark records the highest citation density per capita (about 411.53 citations per million) and high citations per paper, while Switzerland shows similarly strong per-capita performance and a high share of papers crossing major citation thresholds. New Zealand displays the highest publications-per-capita rate (about 7.65 per million), indicating an unusually high national commitment to this topic relative to population.

Middle-income and lower-income countries show lower per-capita output but meaningful presence, often with citation profiles shaped by fewer, highly visible papers and international collaboration patterns. Mexico contributes 87 publications and 1,897 citations with an h-index of 22, placing it within the active contributor group in Latin America; however, its per-capita production and citation density remain below the leading countries, suggesting room to strengthen international visibility and consolidation around highly citable research designs. Brazil and India display substantial output but low per-capita indicators, reflecting the denominator effect of large populations and the need for greater citation concentration to move impact metrics.

Overall, the table indicates a field led by the United States in scale and institutional capacity, while several smaller countries achieve exceptional per-capita impact, and emerging economies contribute growing but less internationally consolidated production.

Table 7.7 summarizes the conceptual structure of the corpus through keyword co-occurrence, and the resulting clusters can be interpreted in direct alignment with the FAO dimensions of food security. The dominance of "food insecurity" and "obesity" at the top of the list is not simply a frequency artifact; it indicates that the field's organizing problem is the co-production of constrained food access with excess weight and related metabolic risk, rather than a narrow focus on undernutrition alone. The simultaneous prominence of "cross-sectional study" and "prevalence" further shows that the mainstream empirical strategy has been surveillance-oriented, privileging population estimates and association mapping over causal identification. This helps explain why risk factor framing is central, with "risk factors," "body mass index," "diabetes mellitus," and "hypertension" anchoring the NCD profile across studies.

In terms of FAO dimensions, "food supply" and "food security" provide the clearest linkage to availability, connecting the literature to upstream food-system conditions and provisioning dynamics. However, the more densely populated cluster aligns with access. Keywords such as "socioeconomics," "poverty," "socioeconomic factors," "health disparity," "health care access," and "food assistance" indicate that much of the research operationalizes food insecurity as a constraint shaped by purchasing power, inequality, and institutional support mechanisms. The presence of "social determinants of health" consolidates this orientation and signals conceptual convergence with broader health-equity scholarship rather than isolated nutrition research.

Utilization is represented by a distinct nutrition-behavior pathway. High-impact keywords such as "nutrition," "diet," "dietary intake," "caloric intake," "nutritional status," "feeding behavior," and "physical activity" reflect the mechanisms through which constrained access translates into diet quality, energy density, and metabolic consequences. The unusually high citation intensity associated with "nutrition" and "diet" indicates that these terms are embedded in widely reused explanatory models and systematic syntheses, reinforced by the strong position of "review."

Stability appears as a thinner but visible cluster, emerging through time-sensitive and shock-related terms. "Longitudinal study" indicates attention to temporal persistence and transitions in food insecurity exposure, while "Coronavirus disease 2019" reflects disruption-driven instability. The inclusion of "depression" and "mental health" suggests that stability is also conceptualized through psychosocial stressors that fluctuate with economic and food-system shocks. Overall, the keyword structure indicates a field heavily weighted toward access and utilization pathways, with

TABLE 7.7
Keyword Co-Occurrence Clusters Aligned with FAO Dimensions

R	Keyword	TP	TC	C/P	≤100	≤10
1	Food insecurity	2,051	57,598	28.1	110	919
2	Obesity	1,720	63,435	36.9	133	992
3	Cross-sectional study	1,563	34,199	21.9	56	730
4	Food security	1,335	41,329	31	88	659
5	Child	1,299	45,189	34.8	102	753
6	Risk factors	994	35,292	35.5	83	571
7	Food supply	875	35,557	40.6	94	573
8	Body mass index	850	24,655	29	53	424
9	Diabetes mellitus	843	32,189	38.2	68	522
10	Prevalence	729	20,767	28.5	46	342
11	Adolescent	685	23,085	33.7	49	366
12	Nutrition	641	55,053	85.9	123	834
13	Diet	627	38,990	62.2	86	612
14	Socioeconomics	506	16,766	33.1	38	269
15	Poverty	495	23,437	47.3	61	345
16	Hypertension	483	14,700	30.4	31	209
17	Nutritional status	476	16,849	35.4	35	237
18	Malnutrition	454	18,341	40.4	34	225
19	Review	430	27,796	64.6	63	295
20	Physical activity	427	13,432	31.5	23	185
21	Feeding behavior	375	13,713	36.6	40	231
22	Social determinants of health	372	9,609	25.8	17	150
23	Socioeconomic factors	366	12,751	34.8	31	220
24	Childhood obesity	343	15,234	44.4	35	198
25	Overweight	320	12,890	40.3	28	178
26	Dietary intake	316	11,567	36.6	25	168
27	Depression	287	8,945	31.2	18	142
28	Longitudinal study	252	9,234	36.6	22	128
29	Food assistance	250	7,890	31.6	15	118
30	Pediatric obesity	238	9,456	39.7	21	132
31	Caloric intake	237	8,934	37.7	19	125
32	Health disparity	232	7,823	33.7	16	112
33	Cardiovascular disease	231	9,567	41.4	22	138
34	Mental health	215	6,789	31.6	14	98
35	Coronavirus disease 2019	205	5,678	27.7	11	87
36	Hunger	193	7,234	37.5	16	102
37	Health care access	193	5,678	29.4	12	87
38	Chronic disease	191	6,890	36.1	15	98
39	Stunting	124	4,567	36.8	10	68
40	Anemia	104	3,890	37.4	9	58

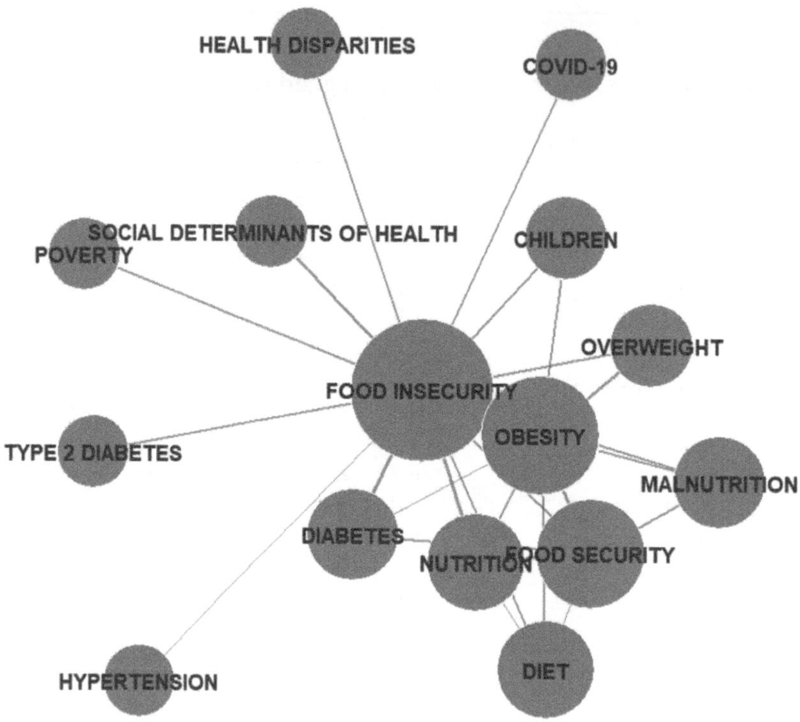

FIGURE 7.2 Keyword co-occurrence map of food insecurity and diet-related non-communicable diseases.

availability and stability present but less densely developed, pointing to clear opportunities for future integration across FAO dimensions.

Figure 7.2 visualizes the author keyword co-occurrence structure of the literature on food insecurity and NCDs. The network is organized around a clear hub, "food insecurity," which occupies the most central position and connects directly to both biomedical outcomes and social-structural determinants. Node size reflects keyword prominence, while edges indicate that concepts are repeatedly mobilized together within the same articles. The topology therefore reveals how the field is framed, not only what it studies (Aria & Cuccurullo, 2017).

Two tightly coupled thematic zones emerge. The first, shown in blue, links food insecurity to the chronic disease profile and equity framing: "obesity," "overweight," "diabetes," "type 2 diabetes," and "hypertension" sit close to the core, indicating that NCD outcomes are not peripheral extensions but primary endpoints in this literature. Surrounding these outcomes are "poverty," "social determinants of health," and "health disparities," which appear as upstream constructs that co-occur with food insecurity and help explain the social gradient in metabolic risk. "Children" is positioned near the hub, suggesting that pediatric populations are a major empirical and policy focus, consistent with the life-course interpretation of early exposure to constrained food access and later cardiometabolic vulnerability.

The second zone, shown in red, consolidates the nutrition and food security mechanism pathway through "diet," "nutrition," "food security," and "malnutrition." Its proximity to the hub indicates that a substantial share of studies interprets food insecurity through utilization mechanisms, particularly diet quality and nutritional status. The co-presence of "obesity" near this red cluster is analytically important: it depicts the field's recurrent emphasis on the double burden logic, where dietary compromise under constrained resources can coexist with excess weight and metabolic disease risk, rather than treating malnutrition and obesity as separate problems.

Finally, the link between "COVID-19" and the central hub signals the incorporation of shock-related instability as a recent driver of food insecurity, but its peripheral placement indicates that this strand is still narrower than the core obesity and social-determinants axis. Overall, the network portrays a field dominated by access and utilization pathways, with NCD outcomes tightly integrated into the conceptual core.

The thematic map provides a compact representation of how the field organizes its dominant conversations by combining relevance (centrality) and internal development (density). Two clusters clearly drive the intellectual agenda. The most consolidated motor theme combines obesity, overweight, children, and malnutrition, indicating that the literature has converged on a life-course framing where early and sustained food-related constraints are studied through weight-related outcomes and nutritional compromise in pediatric populations. This positioning suggests a mature,

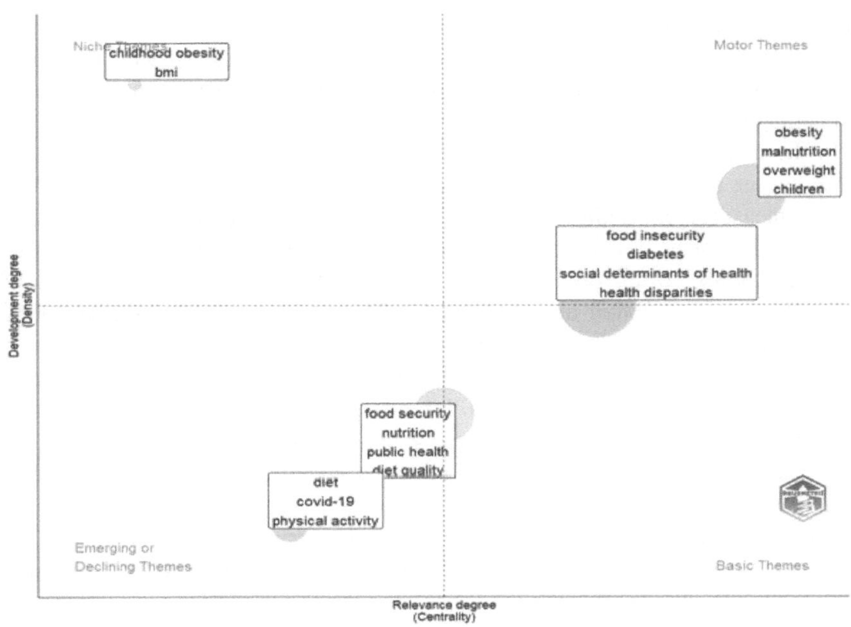

FIGURE 7.3 Strategic thematic map of the food insecurity–non-communicable diseases literature.

internally coherent research front that integrates clinical relevance with population health salience, and it helps explain why obesity-oriented outlets and pediatric venues occupy prominent places in the most cited and most productive segments of the dataset.

A second highly central cluster links food insecurity with diabetes, social determinants of health, and health disparities. Its location close to the core of the map indicates that the field treats food insecurity as more than a dietary descriptor; it functions as an upstream social exposure embedded in equity frameworks and repeatedly connected to metabolic disease outcomes. The presence of social determinants and disparities in the same cluster implies that the dominant explanatory logic in the literature is structural: food insecurity is interpreted through constraints on resources, access, and health-care pathways, rather than through individual choice narratives. Together, these two central clusters show a field that has stabilized around an integrated problem definition: food insecurity is simultaneously a material deprivation, an equity marker, and a clinically consequential risk context.

The basic-theme area contains a broad cluster around food security, nutrition, public health, and diet quality. Its high centrality but lower density is consistent with transversal concepts that appear across many studies as shared framing language rather than as a single tightly bounded specialty. This cluster acts as the connective tissue of the domain, linking work that ranges from dietary pattern measurement to community interventions and policy framing. Meanwhile, the emerging/declining quadrant highlights a smaller cluster combining diet, COVID-19, and physical activity, reflecting a period-sensitive strand where pandemic-related disruption intersects with behavioral mechanisms. Its peripheral placement suggests that this topic is either still consolidating or remains episodic, but it is analytically valuable because it signals how exogenous shocks can redirect research emphasis toward instability and coping behaviors.

In the niche quadrant, the presence of childhood obesity and BMI as a small, dense cluster indicates a technically focused substream, likely driven by pediatric clinical measurement traditions and guideline-oriented research.

Figure 7.4 traces how the field has accumulated knowledge through direct citation links across time, highlighting a relatively coherent intellectual backbone rather than many isolated micro-traditions. The early segment (2001 to 2005) is anchored by papers that established the empirical and conceptual plausibility of the food insecurity and obesity linkage. The appearance of Townsend (2001), Vozoris and Tarasuk (2003), Drewnowski (2004), and Jyoti (2005) signals that the field's formative phase combined two arguments: constrained resources shape diet quality through cost and energy density pathways, and food insecurity is associated with downstream health and child-related outcomes. These early nodes occupy "upstream" positions because later studies repeatedly cite them to justify both measurement choices and the plausibility of paradoxical patterns such as food scarcity coexisting with overweight risk.

From 2007 onward, the network densifies around a recognizable clinical and epidemiologic spine. Seligman (2007) becomes a pivotal connector, linking early conceptual work to later chronic disease evidence, and this role expands in the 2010

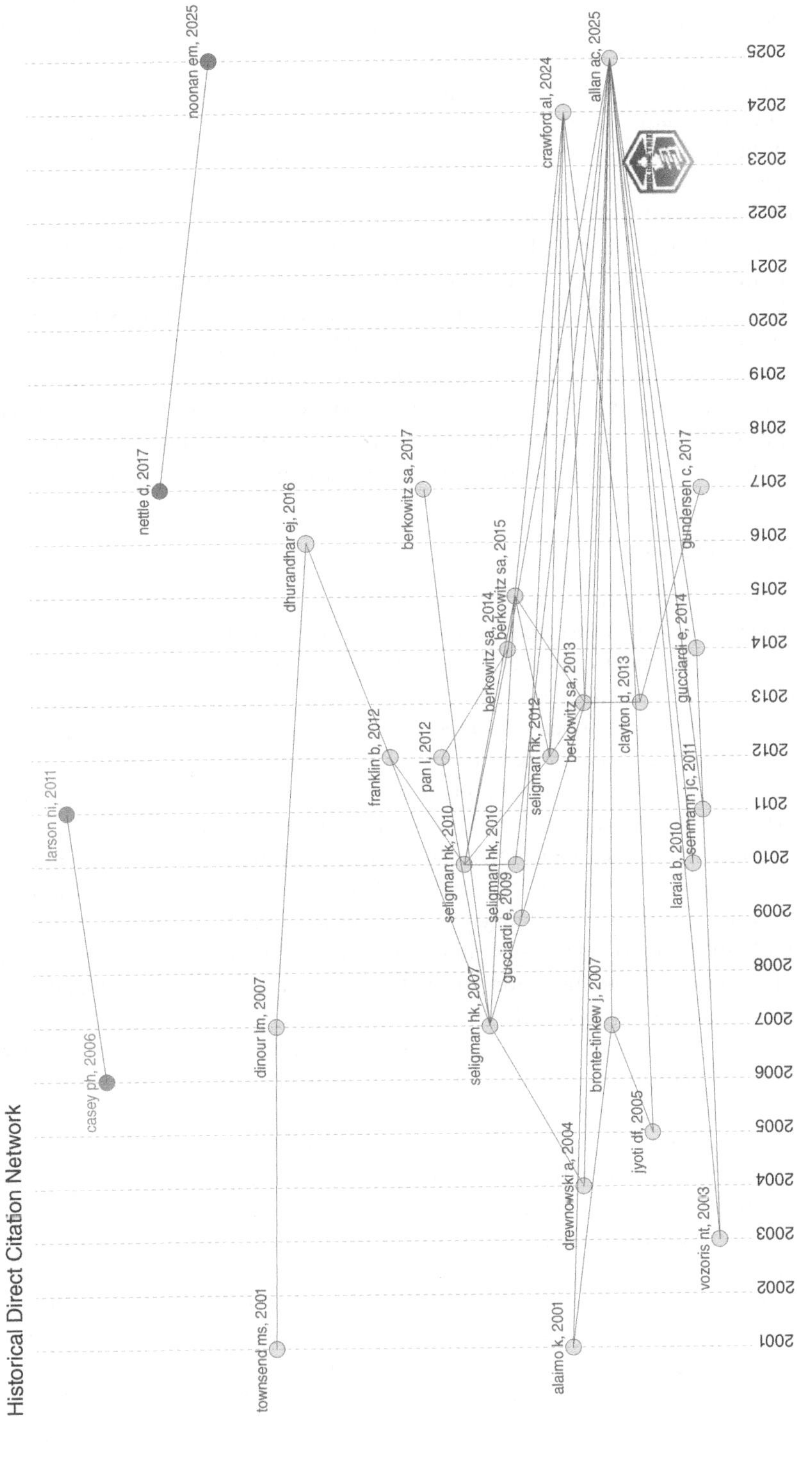

FIGURE 7.4 Cumulative citation network of research on food insecurity and diet-related non-communicable diseases.

to 2014 window through multiple Seligman and Berkowitz nodes. This period corresponds to the consolidation of food insecurity as an exposure variable within large-scale observational designs, particularly those leveraging nationally representative data and clinically interpretable endpoints. Parallel contributions such as Laraia (2010) and Gucciardi, Vogt, DeMelo, & Stewart (2009) form adjacent lines that broaden the field across population groups and care contexts while remaining tied to the central citation core.

A notable feature is the strong continuity from 2012 to 2017, where Franklin et al. (2012), Pan (2012), Clayton (2013), and successive Berkowitz papers (2013, 2014, 2015, 2017) reinforce a shift from documenting associations to clarifying mechanisms and care-relevant implications. The presence of Gundersen (2017) indicates that the field increasingly integrates policy, programmatic, and health services perspectives into the same citation lineage rather than treating them as separate literatures. The network also shows extension into the most recent period, with Crawford (2024) and Allan (2025) positioned as downstream nodes drawing on the earlier core, consistent with a mature research stream where new studies inherit a stable reference base.

7.4 DISCUSSION

The results describe a mature and increasingly policy-relevant research domain in which food insecurity is treated as a social exposure with clinical consequences, rather than as a peripheral socioeconomic descriptor. Across sources, authors, institutions, countries, and thematic structures, the field exhibits a stable core centered on obesity and diabetes outcomes, supported by a strong social-determinants framework and reinforced by pediatric and life-course perspectives. This consolidation is visible in the journal ecosystem (Table 7.2), where high-productivity public health and nutrition outlets provide the main dissemination infrastructure, while a smaller group of clinically oriented journals concentrates disproportionately influential contributions. The coexistence of these editorial roles suggests that the literature has moved beyond an initial exploratory phase and now operates through an established division of labor: broad venues sustain continuous empirical accumulation, whereas high-impact clinical and nutrition journals consolidate integrative evidence that becomes reusable across subfields.

The findings further indicate that research on food insecurity and NCDs has reached a high level of conceptual and empirical maturity, in which food insecurity is no longer understood as a simple socioeconomic marker but as a social exposure with direct clinical implications. The most cited articles and the historical direct citation network (Table 7.3) jointly show that the field's conceptual scaffolding is not based exclusively on food insecurity studies. Instead, it draws extensively on adjacent "anchor" literatures on food systems and climate, maternal and child nutrition interventions, the nutrition transition, and the poverty–obesity relationship. These references provide the explanatory logic through which food insecurity is linked to metabolic risk: resource constraints shape diet quality, dietary compromise can coexist with excess weight, and early-life exposure may influence long-term health trajectories.

This consolidation is further reflected in the stability of the thematic core identified through keyword co-occurrence analysis (Figure 7.2), where obesity and diabetes occupy central positions, closely connected to poverty, inequality, and social determinants of health. The density and persistence of these associations suggest that the field has converged toward a shared interpretation of the metabolic risk associated with food insecurity.

The coexistence of obesity and undernutrition, rather than appearing as a contradiction, emerges as a recurrent pattern within the thematic structure of the field. The clusters identified in the thematic map (Figure 7.3) show that the most influential studies articulate both conditions within the same context of food constraint, characterized by low-cost, low-nutritional-quality diets. This evidence empirically supports the consolidation of the poverty–obesity paradox, widely documented in the most cited documents in the field (Table 7.3), where limited resources do not reduce obesity risk but instead increase it through structural dependence on energy-dense and nutritionally deficient foods.

A second relevant interpretative axis is the centrality of pediatric and life-course perspectives. The thematic evolution of the field (Figure 7.4) reveals a progressive shift from general approaches toward increasing attention to childhood, early-life nutrition, and maternal–child interventions. This trajectory indicates that the literature increasingly recognizes that food insecurity during early life does not have transient effects but rather cumulative ones, shaping long-term metabolic trajectories and increasing future susceptibility to obesity, type 2 diabetes, and hypertension.

Within this framework, diabetes emerges as one of the clinical outcomes most strongly articulated with food insecurity. The direct citation network (Figure 7.4) shows that diabetes-focused studies occupy key structural positions, linking research streams on diet quality, consumption patterns, chronic stress, and irregular access to food. This centrality suggests that the literature interprets the relationship between food insecurity and diabetes as multifactorial and structural, rather than as the result of isolated individual choices.

From an institutional and geographic perspective, patterns of scientific production and leadership reinforce this interpretation. Performance analysis by institutions and countries (Table 7.6) reveals a significant concentration of agenda-setting capacity in academic medical centers and schools of public health in high-income countries, particularly the United States. However, per-capita normalization shows that some smaller countries achieve disproportionate visibility, while emerging economies—despite their high empirical relevance in terms of food insecurity—remain underrepresented in high-impact publication circuits.

Finally, the methodological patterns observed in the analyzed corpus indicate that, although the field is robust in associational terms, it continues to be dominated by observational and cross-sectional designs. This characteristic is reflected in the structure of the methodological clusters identified (Figure 7.3), where prevalence-based and descriptive analyses clearly outweigh longitudinal and quasi-experimental approaches. This evidence points to a clear opportunity to advance toward more integrative analytical models. In particular, the incorporation of intelligent systems, fuzzy logic, and aggregation operators would allow researchers to capture the complexity, uncertainty, and territorial heterogeneity that characterize the relationship between food insecurity and chronic diseases, thereby strengthening both causal inference and the usefulness of evidence for public health decision making.

7.5 CONCLUSION

This study aimed to map, quantify, and interpret the evolution of scientific research linking food insecurity with major NCDs, with a specific emphasis on diabetes, obesity, overweight, hypertension, and closely related determinants. The findings converge on a clear conclusion: the field has reached an advanced level of maturity, consolidating itself as a clinically relevant and policy-relevant line of research, characterized by a stable conceptual core, sustained publication outlets, and clearly identifiable leadership at the levels of authors, institutions, and countries.

From a methodological perspective, the study was conducted using a bibliometric framework that integrated performance analysis with science mapping techniques. Records were retrieved from the Scopus database in January 2026 and restricted to peer-reviewed articles and reviews. The review was based on a robust analytical corpus of 3,650 documents, comprising 3,042 original articles and 608 reviews. The analysis was carried out using the bibliometrix package in R software, combining descriptive and impact metrics for journals, authors, institutions, and countries, along with network-based mapping techniques that enabled the identification of patterns in scientific production, academic impact, and the intellectual and conceptual structure of the field.

The results reveal sustained growth in scientific output since the early 2000s, with a particularly marked acceleration after 2015, coinciding with the global consolidation of food security as a public health priority and a key item on the policy agenda. Bibliometric evidence confirms that food insecurity is closely associated with the development of NCDs, particularly diabetes, obesity, and hypertension. This relationship does not appear to be marginal or incidental but rather constitutes a structural axis of the research field, as reflected in the high frequency and centrality of these terms in keyword co-occurrence maps (see Figure 7.2), where food insecurity is systematically linked to concepts associated with metabolic risk, poverty, diet quality, and health inequalities.

One of the most relevant findings identified in the literature is the consolidation of the so-called poverty–obesity paradox, widely documented in the most highly cited studies in the field (see Table 7.3). These studies demonstrate that food insecurity increases the risk of obesity and diabetes by promoting the consumption of low-cost, low-nutritional-quality foods, characterized by high caloric density, elevated sugar and fat content, and poor micronutrient composition. This pattern is clearly reflected in the dominant thematic clusters identified through the bibliometric analysis (Figure 7.3), where obesity and diabetes are systematically associated with poverty, food insecurity, and unfavorable food environments.

Moreover, the reviewed literature shows that obesity and undernutrition are not mutually exclusive conditions but can coexist within the same individuals and households, particularly in contexts of social vulnerability. This coexistence is recurrent in studies focused on pediatric populations and food-insecure households, as evidenced both in the most influential documents in the field and in the thematic evolution of the research (Figure 7.2). In this regard, the literature consistently indicates that poor nutrition during childhood generates long-term metabolic consequences, increasing susceptibility to obesity, type 2 diabetes, and hypertension in adulthood.

Diabetes emerges as one of the NCDs most strongly articulated with food insecurity, not only through diet quality but also via indirect mechanisms related to chronic stress, irregular access to food, and instability in consumption patterns. This centrality is evident both in the density of publications and in the citation structure of the field (Figure 7.4), where diabetes-focused studies occupy key structural positions and integrate multiple research streams.

At the level of authorship and institutions, scientific leadership is concentrated primarily in researchers and institutions from high-income countries, particularly the United States (Table 7.5), although a growing contribution from middle-income regions has been observed in recent years. This geographic imbalance highlights a significant gap, given that food insecurity and its health consequences disproportionately affect populations in low- and middle-income countries, thereby limiting the contextual diversity of the available evidence.

From a methodological standpoint, the analyzed literature is characterized by the predominance of observational and epidemiological approaches, with still limited integration of adaptive models, decision-support systems, and advanced analytical tools. This finding underscores the need to move toward more integrative approaches capable of capturing the complexity, uncertainty, and territorial heterogeneity inherent in the relationship between food insecurity and chronic diseases.

Nevertheless, these findings should be interpreted in light of certain limitations. Reliance on a single database may introduce coverage and language biases, which are common in bibliometric studies. Additionally, citation-based indicators involve time-lag effects that may underrepresent recent contributions not yet consolidated in terms of impact. Finally, the predominance of observational studies limits the ability to establish direct causal inferences.

In this context, future research should incorporate longitudinal designs, quasi-experimental strategies, and adaptive models to strengthen causal inference and enhance the translation of evidence into more effective public policies. In particular, the integration of intelligent systems, fuzzy logic, and decision-support tools represents a promising avenue for improving understanding of underlying mechanisms and guiding more precise interventions.

From an applied perspective, the findings reinforce the need for policies and interventions that go beyond ensuring access to food and instead structurally address diet quality, especially among vulnerable populations and during early stages of the life course. Addressing food insecurity in an integrated manner is essential to reducing the burden of NCDs and advancing toward more equitable, sustainable, and socially just health and food systems.

REFERENCES

Allan, A. C., Buxton, O. M., Ross, L. A., Engeland, C. G., Thorpe, R. J., Beydoun, M. A., Zonderman, A. B., Evans, M. K., & Gamaldo, A. A. (2025). HANDLSleep: Associations among neighborhood quality, food security, and cognitive function. *Alzheimer's & Dementia*, *21*(S6), e098034. https://doi.org/10.1002/alz70860_098034

Allen, L. N., & Feigl, A. B. (2017). Reframing non-communicable diseases as socially transmitted conditions. *The Lancet Global Health*, *5*(7), e644–e646. https://doi.org/10.1016/S2214-109X(17)30200-0

Aria, M., & Cuccurullo, C. (2017). bibliometrix: An R-tool for comprehensive science mapping analysis. *Journal of Informetrics, 11*(4), 959–975. https://doi.org/10.1016/j.joi.2017.08.007

Bhutta, Z. A., Das, J. K., Rizvi, A., Gaffey, M. F., Walker, N., Horton, S., … Black, R. E. (2013). Evidence-based interventions for improvement of maternal and child nutrition: What can be done and at what cost? *The Lancet, 382*(9890), 452–477. https://doi.org/10.1016/S0140-6736(13)60996-4

Blanco-Mesa, F., León-Castro, E., & Merigó, J. M. (2019, August 1). A bibliometric analysis of aggregation operators. *Applied Soft Computing Journal*, Vol. 81. https://doi.org/10.1016/j.asoc.2019.105488

Blondel, V. D., Guillaume, J.-L., Lambiotte, R., & Lefebvre, E. (2008). Fast unfolding of communities in large networks. *Journal of Statistical Mechanics: Theory and Experiment, 2008*(10), P10008. https://doi.org/10.1088/1742-5468/2008/10/P10008

Bornmann, L., & Leydesdorff, L. (2014). Scientometrics in a changing research landscape: bibliometrics has become an integral part of research quality evaluation and has been changing the practice of research. *EMBO Reports, 15*(12), 1228–1232.

Burrell, Q. L. (2002). The nth-citation distribution and obsolescence. *Scientometrics, 53*(3), 309–323.

Callon, M., Courtial, J. P., & Laville, F. (1991). Co-word analysis as a tool for describing the network of interactions between basic and technological research: The case of polymer chemistry. *Scientometrics, 22*(1), 155–205.

Crawford, K., Kerr, W., & Tyl, C. (2024). Effect of hydrocolloid addition on cake prepared with aquafaba as egg substitute. *International Journal of Food Science & Technology, 59*(1), 552–559. https://doi.org/10.1111/ijfs.16492

Darmon, N., & Drewnowski, A. (2015). Contribution of food prices and diet cost to socioeconomic disparities in diet quality and health: A systematic review and analysis. *Nutrition Reviews, 73*(10), 643–660.

Dietz, W. H. (1995). Does hunger cause obesity? *Pediatrics, 95*(5), 766–767.

Dinour, L. M., Bergen, D., & Yeh, M.-C. (2007). The food insecurity-obesity paradox: A review of the literature and the role food stamps may play. *Journal of the American Dietetic Association, 107*(11), 1952–1961. https://doi.org/10.1016/j.jada.2007.08.006

Donthu, N., Kumar, S., Mukherjee, D., Pandey, N., & Lim, W. M. (2021). How to conduct a bibliometric analysis: An overview and guidelines. *Journal of Business Research, 133*, 285–296.

Drewnowski, A., & Popkin, B. M. (1997). The nutrition transition: New trends in the global diet. *Nutrition Reviews, 55*(2), 31–43. https://doi.org/10.1111/j.1753-4887.1997.tb01593.x

Drewnowski, A. (2004). Obesity and the food environment. *American Journal of Preventive Medicine, 27*(3), 154–162. https://doi.org/10.1016/j.amepre.2004.06.011

Drewnowski, A., & Specter, S. E. (2004). Poverty and obesity: The role of energy density and energy costs. *American Journal of Clinical Nutrition, 79*(1), 6–16. https://doi.org/10.1093/ajcn/79.1.6

Food and Agriculture Organization of the United Nations [FAO]. (2001). *The State of Food Insecurity in the World 2001.* Retrieved from https://openknowledge.fao.org/server/api/core/bitstreams/73b3133d-b1a6-439a-b0b8-e0adf3aaa7d9/content

Franklin, B., Jones, A., Love, D., Puckett, S., Macklin, J., & White-Means, S. (2012). Exploring mediators of food insecurity and obesity: A review of recent literature. *Journal of Community Health, 37*(1), 253–264.

Fruchterman, T. M. J., & Reingold, E. M. (1991). Graph drawing by force-directed placement. *Software: Practice and Experience, 21*(11), 1129–1164.

García-Orozco, D., Espitia-Moreno, I. C., Alfaro-García, V. G., & Merigó, J. M. (2020). Sustainability in Mexico a bibliometric analysis of the scientific research field presented in the last 28 years. *Inquietud Empresarial, 20*(2), 101–120.

García-Orozco, D., Alfaro-García, V. G., Merigó, J. M., Espitia Moreno, I. C., & Gómez Monge, R. (2022). An overview of the most influential journals in fuzzy systems research. *Expert Systems with Applications*, *200*, 117090. https://doi.org/10.1016/j.eswa.2022.117090

Garfield, E. (1972). Citation analysis as a tool in journal evaluation: Journals can be ranked by frequency and impact of citations for science policy studies. *Science*, *178*(4060), 471–479.

Grant, M. J., & Booth, A. (2009). A typology of reviews: An analysis of 14 review types and associated methodologies. *Health Information & Libraries Journal*, *26*(2), 91–108.

Gucciardi, E., Vogt, J. A., DeMelo, M., & Stewart, D. E. (2009). Exploration of the relationship between household food insecurity and diabetes in Canada. *Diabetes Care*, *32*(12), 2218–2224.

Gundersen, C., Kreider, B., & Pepper, J. V. (2017). Partial identification methods for evaluating food assistance programs: a case study of the causal impact of snap on food insecurity. *American Journal of Agricultural Economics*, *99*(4), 875–893. https://doi.org/10.1093/ajae/aax026

Hampl, S. E., Hassink, S. G., Skinner, A. C., Armstrong, S. C., Barlow, S. E., Bolling, C. F., … Joseph, M. M. (2023). Clinical practice guideline for the evaluation and treatment of children and adolescents with obesity. *Pediatrics*, *151*(2).

Health, H. (2003). Household food insufficiency is associated with poorer health. *The Journal of Nutrition*, *133*(1), 120–126.

Hill-Briggs, F., Adler, N. E., Berkowitz, S. A., Chin, M. H., Gary-Webb, T. L., Navas-Acien, A., … Haire-Joshu, D. (2020). Social determinants of health and diabetes: A scientific review. *Diabetes Care*, *44*(1), 258.

Hirsch, J. E. (2005). An index to quantify an individual's scientific research output. *Proceedings of the National Academy of Sciences*, *102*(46), 16569–16572.

Jyoti, D. F., Frongillo, E. A., & Jones, S. J. (2005). Food insecurity affects school children's academic performance, weight gain, and social skills. *The Journal of Nutrition*, *135*(12), 2831–2839.

Laraia, B. A., Siega-Riz, A. M., & Gundersen, C. (2010). Household food insecurity is associated with self-reported pregravid weight status, gestational weight gain, and pregnancy complications. *Journal of the American Dietetic Association*, *110*(5), 692–701. https://doi.org/10.1016/j.jada.2010.02.014

Merigó, J. M., & Yang, J.-B. (2017). A bibliometric analysis of operations research and management science. *Omega*, *73*, 37–48.

Monge, R. G., Alfaro-García, V. G., Espitia-Moreno, I. C., García-Orozco, D., & de Vivar Mercadillo, M. R. R. (2022). Environmental sustainability: A 10-year bibliometric analysis of the developments in management. *Business, Finance and Economics*. https://doi.org/10.1007/978-3-030-96150-3_18

Mongeon, P., & Paul-Hus, A. (2016). The journal coverage of Web of Science and Scopus: A comparative analysis. *Scientometrics*, *106*(1), 213–228.

Newman, M. E. J. (2001). The structure of scientific collaboration networks. *Proceedings of the National Academy of Sciences*, *98*(2), 404–409.

Phelps, N. H., Singleton, R. K., Zhou, B., Heap, R. A., Mishra, A., Bennett, J. E., … Ezzati, M. (2024). Worldwide trends in underweight and obesity from 1990 to 2022: A pooled analysis of 3663 population-representative studies with 222 million children, adolescents, and adults. *The Lancet*, *403*(10431), 1027–1050. https://doi.org/10.1016/S0140-6736(23)02750-2

Popkin, B. M. (1999). Urbanization, lifestyle changes and the nutrition transition. *World Development*, *27*(11), 1905–1916.

Popkin, B. M. (2002). The shift in stages of the nutrition transition in the developing world differs from past experiences! *Public Health Nutrition*, *5*(1A), 205–214.

Price, D. J. D. S. (1965). Networks of scientific papers: The pattern of bibliographic references indicates the nature of the scientific research front. *Science*, *149*(3683), 510–515.

Seligman, H. K., Bindman, A. B., Vittinghoff, E., Kanaya, A. M., & Kushel, M. B. (2007). Food insecurity is associated with diabetes mellitus: Results from the National Health Examination and Nutrition Examination Survey (NHANES) 1999–2002. *Journal of General Internal Medicine*, *22*(7), 1018–1023. https://doi.org/10.1007/s11606-007-0192-6

Seligman, H. K., Bindman, A. B., Vittinghoff, E., Kanaya, A. M., & Kushel, M. B. (2007). Food insecurity is associated with diabetes mellitus: results from the national health examination and nutrition examination survey (NHANES) 1999–2002. *Journal of General Internal Medicine*, *22*(7), 1018–1023. https://doi.org/10.1007/s11606-007-0192-6

Seligman, H. K., Laraia, B. A., & Kushel, M. B. (2010). Food insecurity is associated with chronic disease among low-income NHANES participants. *Journal of Nutrition*, *140*(2), 304–310. https://doi.org/10.3945/jn.109.112573

Small, H. (1973). Co-citation in the scientific literature: A new measure of the relationship between two documents. *Journal of the American Society for Information Science*, *24*(4), 265–269.

Townsend, M. S., Peerson, J., Love, B., Achterberg, C., & Murphy, S. P. (2001). Food insecurity is positively related to overweight in women. *The Journal of Nutrition*, *131*(6), 1738–1745.

van Eck, N. J., & Waltman, L. (2010). Software survey: VOSviewer, a computer program for bibliometric mapping. *Scientometrics*, *84*(2), 523–538.

Vozoris, N. T., & Tarasuk, V. S. (2003). Household food insufficiency is associated with poorer health. *The Journal of Nutrition*, *133*(1), 120–126. https://doi.org/10.1093/jn/133.1.120

Wheeler, T., & Von Braun, J. (2013). Climate change impacts on global food security. *Science*, *341*(6145), 508–513. https://doi.org/10.1126/science.1239402

WHO. (2023). Noncommunicable diseases. Retrieved from World Health Organization's website: https://www.who.int/news-room/fact-sheets/detail/noncommunicable-diseases

World Health Organization's [WHO]. (2022). *World Health Organization 2022*. Retrieved from https://iris.who.int/server/api/core/bitstreams/2cace9cd-cd8f-4b68-9d57-657dc5558f32/content

Yepes-Nuñez, J. J., Urrútia, G., Romero-García, M., & Alonso-Fernández, S. (2021). The PRISMA 2020 statement: An updated guideline for reporting systematic reviews Declaración PRISMA 2020: *una guía actualizada para la publicación de revisiones sistemáticas. Revista Espanola de Cardiologia*, *74*(9), 790–799.

Zhao, J., & Li, M. (2023). Worldwide trends in prediabetes from 1985 to 2022: A bibliometric analysis using bibliometrix R-tool. *Frontiers in Public Health*, *11*, 1072521.

8 Smart Tourism Destinations

A Local Assessment Model for ICT Indicators

Ana Laura Herrera-Prado, Mónica Velarde-Valdez, Elizabeth Olmos-Martínez, and Margarita Guadalupe Zazueta-Hernández

8.1 INTRODUCTION

In the past decade, the accelerated integration of information and communication technologies (ICTs) across all spheres of social life has extended to the practices and operations of tourism sector stakeholders. The intensification of this trend was particularly evident during the period of physical distancing measures mandated in response to the COVID-19 pandemic. As a result, tourist destinations are formulating and deploying ICT-driven innovation strategies to align with the expectations and digital engagement preferences characteristic of the smart tourist paradigm (Herrera-Prado et al., 2024a; Ivars-Baidal et al., 2019, 2021; Montecé-Macías and Marcial-Chang, 2023; SEGITTUR, 2025). Various forms of tourism management, commonly referred to as tourism destination models, have been developed worldwide with the objective of enhancing the competitiveness of destinations and equipping them to respond to the evolving challenges posed by economic, social, and technological transformations (Herrera-Prado, 2021; Iglesias et al., 2018; Ivars-Baidal et al., 2017).

One category of tourism destination management models that emphasizes the integration of ICT is the smart tourism destination (STD) framework. In alignment with this global trend, numerous STD models have been formulated over the past decade (Cabral, 2018; Cimbaljević et al., 2018; EU, 2019; Invattur, 2015; Ivars-Baidal et al., 2018; SEGITTUR, 2018). In 2012, the STD model proposed by the Secretary of State for Tourism of Spain was developed by the State Society for the Management of Innovation and Tourist Technologies (Sociedad Estatal para la Gestión de la Innovación y las Tecnologías Turísticas, SEGITTUR), in collaboration with the Valencian Institute of Tourist Technologies. STDs are characterized as innovative environments supported by advanced technological infrastructure, designed to foster sustainable development, ensure accessibility, enhance visitor interaction with their surroundings, improve the quality of tourist experiences, and contribute to residents' quality of life (SEGITTUR, 2025).

DOI: 10.1201/9781003724902-9

The structure of STD models is composed of strategic axes and their corresponding indicators, typically integrated into graphical representations that constitute the core of each model. These components provide the basis for implementing continuous improvement processes in the management of tourist destinations, enabling adaptation to current and future challenges (Cerdá-Mansilla et al., 2024; Invattur, 2015; Ivars-Baidal et al., 2016, 2017; López de Ávila et al., 2015; SEGITTUR, 2025). STD certification promotes a management approach that recognizes the cross-cutting nature of tourism activity and the distinctive characteristics of each destination. To obtain this distinction, destinations undergo a diagnostic evaluation that yields a set of recommendations, an action plan, and a monitoring mechanism.

However, the proliferation of STD model proposals has generated conceptual redundancy, creating uncertainty for destinations seeking certification. Addressing this gap, this chapter introduces the design of the assessment model for ICT indicators (AMII). This model consolidates and refines indicators from prevailing STD frameworks, offering a standardized and innovative local assessment tool to evaluate ICT availability and utilization in destinations pursuing STD recognition, regardless of their chosen model.

8.2 THEORETICAL FRAMEWORK

Smart tourism is an adaptation of the concept of smart cities applied to the tourism domain and constitutes a management model that employs technology as the primary tool to achieve efficient and sustainable development in destinations (Moliner-Tena et al., 2025). This model focuses on the use of ICTs, the Internet of Things (IoT), Big Data, artificial intelligence, and digital platforms to benefit residents and businesses, optimize visitor experiences, and manage resources sustainability (Alsharif et al., 2024; Carballido & Guevara-Plaza, 2021; Gretzel et al., 2015). STDs are thus conceived as the application of smart city principles within an interconnected system of tourism services that respond to the needs of visitors, residents, and tourism operators.

According to UNWTO (2025), *"a smart destination is one with a strategy of technology, innovation, sustainability, accessibility, and inclusion throughout the entire tourism cycle: before, during, and after the trip."* This definition underscores the integration of such strategies across all phases of travel, consolidating STD as key actors in a new tourism paradigm.

One of the most widely cited definitions of STD is provided by SEGITTUR, which states that a smart destination is

> an innovative tourist destination, built upon cutting-edge technological infrastructure, ensuring the sustainable development of the tourism territory, accessible to all, facilitating visitor interaction and integration with the environment, while enhancing the quality of the visitor experience and improving residents' quality of life.
>
> **(SEGITTUR, 2025)**

In this regard, STDs aim to improve visitor interaction with destinations by prioritizing quality and incorporating tourist demands and contextual characteristics into the planning of routes, itineraries, and recommendations (Alsharif et al., 2024; Wang, 2024). Technology is therefore embedded in diverse activities and sites chosen by tourists, from the motivation and planning stages of the trip to the on-site experience (Herrera-Prado et al., 2024b), in response to new requirements posed by both tourists and operators. Overall, these dynamics underscore the central role of technology in shaping visitor interaction and destination management across the tourism experience.

The intensive use of ICT, such as digital platforms, social media, and mobile applications, combined with technological infrastructure and collaborative governance, represents the core features of STDs. These elements aim to promote sustainable development, enhance destination competitiveness, improve residents' quality of life, and optimize tourist experiences. Together, these components form an integrated framework that underpins the strategic planning and management of STDs.

Over the past decades, the scientific literature has proposed various conceptual STD models (often represented graphically) for effective, holistic, and systemic planning and management (Cerdá-Mansilla et al., 2024; Herrera-Prado et al., 2024a; Ivars-Baidal et al., 2018). These models vary in scope and focus, sometimes generating confusion or ambiguity (Cerdá-Mansilla et al., 2024), particularly in relation to destination management and operational practices. As a result, the growing diversity of STD models has made conceptual alignment within the field increasingly challenging. These models are generally structured around strategic axes, each containing a set of indicators whose interrelations reflect the diversity of factors involved in destination development and management.

The STD model proposed by Invattur (2015) emphasizes ICT implementation as a core element, integrating indicators related to connectivity, information systems, innovation, governance, and sustainability. Similarly, Ivars-Baidal et al. (2018) propose a systemic representation of STD through a three-level management approach: (1) strategic-relational, including governance, sustainability, and innovation; (2) instrumental, encompassing connectivity, sensorization, and information systems; and (3) application, involving smart solutions. Regarding competitiveness, Cimbaljević et al. (2018) identify resources, major attractions, supporting factors, smart mobility, smart governance, ICT, and tourism ecosystem actors as components of STD models aimed at enhancing destination competitiveness. Cabral (2018) further emphasizes cooperation among model components to strengthen competitiveness, highlighting comparative advantages and competitive advantages, alongside governance, ICT, and information systems.

Additionally, the proposal made by SEGITTUR through the Smart Destination Programme in Spain is noteworthy. This initiative seeks to enhance the competitiveness of tourist destinations internationally while contributing to residents' quality of life. According to SEGITTUR's methodology (2022), STDs are structured around five fundamental pillars: (1) governance, (2) accessibility, (3) sustainability, (4) technology, and (5) innovation. Governance establishes the overarching framework for efficient STD implementation by fostering cooperation among institutions and stakeholders, guided by principles of transparency and citizen participation, and supporting strategic decision-making and management. Accessibility ensures that services, spaces, and platforms are inclusive and universal, promoting visitor-resident interaction and integration, thereby enhancing visitor experiences and resident benefits. Sustainability seeks to balance tourism development with environmental and sociocultural preservation, prioritizing resource efficiency and impact reduction across all STD dimensions. Technology involves the efficient use of ICT such as IoT, AI, digital platforms, mobile devices, and big data systems to provide accessibility, information, personalization, and management efficiency. Innovation fosters novel products and services supported by technological infrastructure and sociocultural well-being, integrating visitor experiences and promoting value co-creation, thereby differentiating destinations and enabling collaborative branding (Cerdá-Mansilla et al., 2024; Gretzel et al., 2015; SEGITTUR, 2022). The pillars structuring STD consolidate the incorporation of key dimensions such as competitiveness, co-creation, and cooperation, alongside sustainability and technology, as essential components for STD implementation and management through strategic planning, ultimately aiming for successful tourism development in destinations.

The origins of STDs, as well as definitions published in specialized literature, underscore the importance of modern technologies in contemporary tourism. Moreover, the definitions presented in the literature converge on the strategic incorporation of ICTs as a central element of this concept. Such technologies are understood not only as tools for digitization, but also as means of improving destination management, optimizing decision-making, enhancing competitiveness, promoting sustainability, and enriching the tourism experience, while also addressing residents' needs. This theoretical framework shows how ICTs are integrated into STDs through conceptual models that articulate strategic axes such as governance, innovation, accessibility, sustainability, and technology. However, it also reveals a variety of proposed models developed from institutional, academic, and territorial approaches. This multiplicity of proposed models could lead to conceptual confusion and ambiguity regarding the structure, scope, and relationships among the components of STD models, which hinders their comparison and consistent application. Nevertheless, the analysis of these models could allow the identification of common structural elements that provide a basis for understanding the operational logic of ICT use in STDs and support the development of an AMII.

8.3 METHODOLOGY

For the development of the AMII, a mixed-methods strategy was employed, following the research methodology guidelines proposed by Hernández-Sampieri et al. (2014), and based on an exploratory, analytical, and theoretical evaluation of existing STD conceptual models. A total of six models were initially identified; however, only four were selected for structural and content analysis due to the accessibility of detailed indicator information. Models lacking publicly available data were therefore excluded from the study.

Through a structural analysis of the examined STD models, their strategic axes were compared to derive a set of common axes for the AMII. Using the presence of ICT-related indicators as the basis for categorization, a content analysis was conducted for each model, resulting in two indicator sets: one labeled "ICT Indicators"

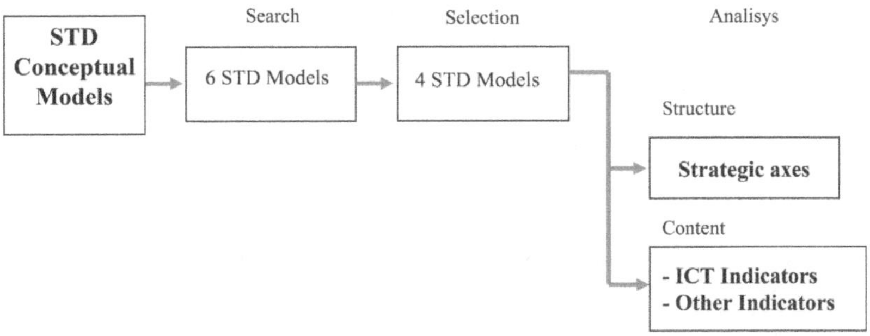

FIGURE 8.1 Structural analysis and categorization of indicators derived from conceptual smart tourism destination (STD) models. ICT = information and communication technology.

Source: Authors' own elaboration.

and the other "Other Indicators" (Figure 8.1). Subsequently, a refinement process was applied, during which duplicate or analogous indicators were discarded to enhance coherence and eliminate redundancy.

To support the application of the analytic hierarchy process (AHP) in selecting ICT indicators and designing the AMII, a closed-ended questionnaire was developed using the Google Forms® platform, consisting of two sections. The first section aimed to assess the expert panel members' weighting of each proposed strategic axis. While the second section assessed the perceived relevance of the identified ICT indicators.

Regarding the structure of the questionnaire, the first section presented the strategic axes proposed for a simplified structure based on the STD model for the AMII. Expert panel members were asked to weigh the importance of each axis by assigning values from 1 (not important) to 5 (very important). In the second section, they assessed the relevance of each ICT indicator using a scale from 0 (not relevant) to 5 (completely relevant).

This instrument was applied to a panel of academic experts, following the methodology proposed by Cabral (2018). The objective was to collect expert judgments on the weighting of each strategic axis within the context of developing a model based on ICT indicators for the self-assessment of tourist destinations. Experts also evaluated the relevance of ICT indicators that could be used to measure their availability and utilization within tourist destinations.

The expert panel was selected from a list of potential participants according to their academic qualifications. Specifically, candidates were required to be affiliated with higher education institutions and to demonstrate substantial research experience in the fields of tourism and ICTs. Ultimately, the panel comprised ten academic experts.

The AHP was employed for the final selection of ICT indicators and the development of their measurement scales (as well as for determining the relative importance of each strategic axis), following the methodology of Saaty (1987), Wang et al. (2016), and Kontogianni et al. (2018) (Figure 8.2). The AHP process involved the following key steps, adapted from Berges-García et al. (2016): (1) establishing

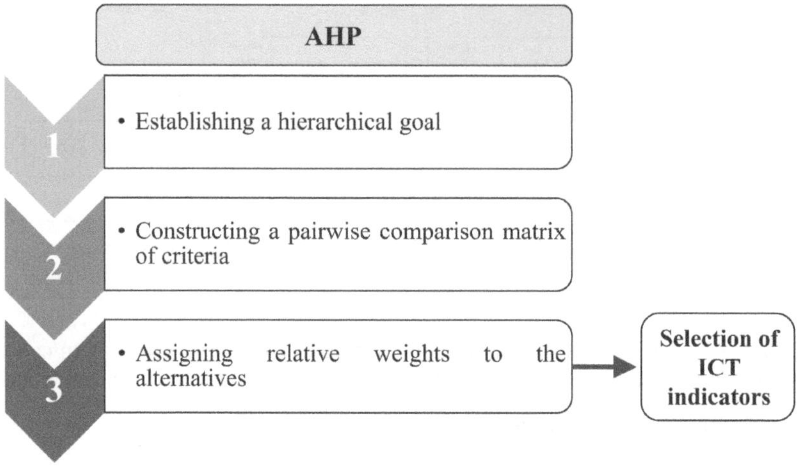

FIGURE 8.2 Application of the analytic hierarchy process (AHP) to select ICT indicators.

Source: Adapted by the authors based on Saaty (1987), Berges-García et al. (2016), Wang et al. (2016), and Kontogianni et al. (2018).

a hierarchical goal, (2) constructing a pairwise comparison matrix of criteria, and (3) assigning relative weights to the alternatives.

Scilab® software was used to perform the necessary calculations to obtain the global matrix resulting from the pairwise comparison of expert responses. This matrix was then processed using Microsoft Excel®, for additional calculations to generate the priority vectors for each group of indicators, organized according to their corresponding strategic axis.

Finally, the AHP is one of the most widely used techniques for multi-criteria decision-making. In this study, AHP provided a formal structure for the development of the AMII by integrating both qualitative and quantitative criteria derived from expert panel input, thereby reducing potential subjectivity and enhancing methodological robustness.

8.4 RESULTS

8.4.1 Structural Analysis

A preliminary analysis of the identified models indicated that, among the various STD models described in the scientific literature, six are particularly relevant for consideration: the *SEGITTUR* model, the *Invattur* model (Comunitat Valenciana), the *Competitive* model, the *Systemic* model, the *Synergic* model, and the *European Union* model. These models share structural similarities, notably their organization into strategic axes and indicators. Of these six models, four were selected for detailed analysis based on the availability of structural information: *Invattur, Competitive, Systemic,* and *Synergic*. The indicators associated with these four models were classified into two categories: those related to ICTs (ICT indicators) and those unrelated to technological tools (other indicators). Table 8.1 presents the number of strategic

TABLE 8.1
Distribution and Percentage of ICT Indicators across Selected Smart Tourism Destination (STD) Models

STD Models	Strategic Axes	Number of Indicators	Number of ICT Indicators	Percentage of ICT Indicators
Invattur	8	78	47	60%
Competitive	4	205	68	33%
Systemic	8	62	47	76%
Synergic	8	88	28	32%

Source: Compiled and analyzed by the authors.

axes, the total number of indicators, and the number and percentage of ICT indicators for each selected STD model.

The *Systemic* model (Ivars-Baidal et al., 2018) includes the highest proportion of ICT indicators, with 76% of its total indicators dedicated to evaluating ICT aspects of STD. The *Invattur* model (Invattur, 2015) follows with 60%, in contrast, the *Competitive* model (Cimbaljević et al., 2018) and the *Synergic* model (Cabral, 2018) allocate a lower proportion of ICT indicators, at 33% and 32%, respectively. These results indicate substantial variability in the emphasis placed on ICT across the selected STD models. Additionally, the review identified three STD models (*SEGITTUR, Invattur,* and the *European Union* model) that have been implemented in various tourist destinations worldwide.

Based on an analysis of the structure of the selected STD models, a list of identified ICT indicators is presented in Annexes 8.1–8.4. Annex 8.1 presents the structure of the *Invattur* model, which comprises eight strategic axes: Accessibility, Connectivity and Sensorization, Governance, Tourist Information, Innovation, Online Marketing, Information Systems/Smart Tourism, and Territorial Sustainability. In total, 47 ICT indicators were identified in this model.

Annex 8.2 lists 68 ICT indicators identified in the *Competitive* model, distributed across four strategic axes: Competitiveness, Tourism Components, Smart, and Digital Services. Annex 8.3 presents the eight strategic axes defined in the *Systemic* model: Accessibility, Connectivity and Sensorization, Governance, Tourist Information, Innovation, Online Marketing, Information Systems, and Sustainability. In total, 47 ICT indicators were identified in the *Systemic* model.

Finally, Annex 8.4 presents 28 ICT indicators identified in the *Synergic* model, organized across eight strategic axes: Tourist Attractions and Supply Conditions; Competitive and Social Conditioning of the Urban Environment; Demand and Economic Impact; Governance and Strategic Management; Innovation; Sustainability and Environment; ICT, Marketing, and Information Systems; and Accessible Tourism.

The comparative analysis of these model structures served as the basis for proposing a common set of strategic axes for the AMII. As a result, the ICT indicators were grouped into five strategic axes: Governance, Accessibility, Sustainability,

Innovation, and Technology. This grouping supports the development of a new evaluation structure applicable to tourist destinations seeking STD certification. The four analyzed STD models contained 433 indicators in total, of which 190 were classified as ICT indicators (see Annexes 8.1–8.4). A refinement process was conducted to eliminate repeated, analogous, or redundant indicators across models. This process resulted in a final set of 72 unique ICT indicators.

Based on the information obtained, a structured questionnaire was developed and administered to the expert panel. This instrument incorporated a preliminary version of the proposed AMII, consisting of the following strategic axes and associated ICT indicators: Governance (9 indicators), Accessibility (18), Sustainability (8), Innovation (6), and Technology (31). The objective of this research phase was to identify the ICT indicators that academic experts and researchers in tourism and ICTs considered most relevant for assessing the degree of ICT utilization in tourist destinations.

8.4.2 Analytic Hierarchy Process

According to the responses provided by the expert panel, the AHP was used to identify and prioritize the ICT indicators considered most relevant for the evaluation of tourist destinations within the AMII. The results corresponding to each step of the process are presented below.

1. **Establishment of a Hierarchical Goal for ICT indicators.** To initiate the AHP process, a decision hierarchy tree was developed for each strategic axis: Governance, Accessibility, Sustainability, Innovation, and Technology. By way of illustration, Figure 8.3 presents the decision hierarchy tree corresponding to the Governance axis. The first level defines the

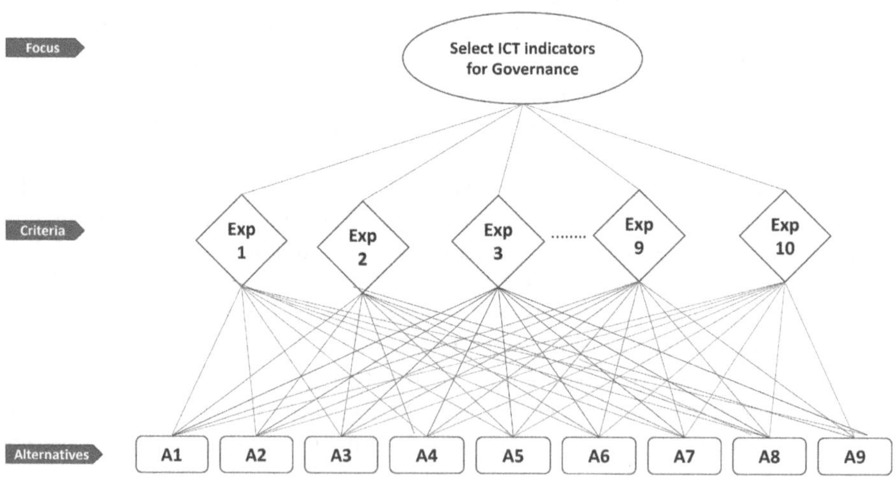

FIGURE 8.3 Decision hierarchy tree for ICT indicators in the Governance axis.

Source: Compiled by the authors based on Saaty (1987) and Kontogianni et al. (2018).

overall objective: to select the most relevant ICT indicators for the respective axis. The second level represents the selection criterion, which in this case is based on the opinions expressed by the academic experts (Expl-Exp10). The third level lists all potential ICT indicator alternatives (A1-A9) proposed for the Governance axis.

2. **Construction of the Pairwise Comparison Matrix.** Continuing with the Governance axis as an example, ten pairwise comparison matrices were constructed, one for each expert on the panel. Each matrix compared the proposed ICT indicators against one another within the same strategic axis. These matrices were then consolidated into a single global matrix, integrating the judgments from all ten experts regarding the ICT indicators for the strategic axis under analysis. For the Governance axis, this process resulted in a 9 × 9 matrix, corresponding to the nine ICT indicators under analysis (Table 8.2). The same procedure was applied to each of the five strategic axes.
3. **Assignment of relative importance (Priority Analysis).** The relative importance of the ICT indicators was calculated for each strategic axis using the AHP methodology. Using the Governance axis as an example, a normalized matrix (Table 8.3) was derived from the global matrix (Table 8.2). From this normalized matrix, the average vector (also known as the priority vector) was calculated. This vector contains the weights assigned to each ICT indicator, reflecting its relative priority according to expert judgment.

These weights were then used to rank the indicators, as shown in Table 8.4. The same procedure was applied to the remaining strategic axes of the proposed AMII. To determine which indicators would be retained, only those contributing to a cumulative weight of up to 0.80 were selected, following the criterion suggested by Berges-García et al. (2016). In the case of the Governance axis, this resulted in

TABLE 8.2
Consolidated Pairwise Comparison Matrix for Governance Axis Indicators

	A1	A2	A3	A4	A5	A6	A7	A8	A9
A1	1.00	1.84	4.83	1.72	2.37	2.67	2.14	3.94	1.25
A2	1.00	1.00	2.26	0.80	1.11	1.25	1.00	1.84	0.68
A3	0.54	1.00	1.00	0.27	0.37	0.44	0.34	0.64	0.23
A4	0.21	0.44	1.00	1.00	1.55	1.75	1.25	3.00	0.72
A5	0.58	1.25	3.68	1.00	1.00	1.12	0.80	1.93	0.53
A6	0.42	0.90	2.67	0.64	1.00	1.00	0.80	1.48	0.47
A7	0.37	0.80	2.29	0.57	0.89	1.00	1.00	2.14	0.58
A8	0.47	1.00	2.95	0.80	1.25	1.25	1.00	1.00	0.32
A9	0.25	0.54	1.55	0.33	0.52	0.68	0.47	1.00	1.00

Source: Created by the authors.

TABLE 8.3
Normalized Matrix and Priority Vector for Governance Indicators

	A1	A2	A3	A4	A5	A6	A7	A8	A9
A1	0.21	0.21	0.22	0.24	0.24	0.24	0.24	0.23	0.22
A2	0.21	0.11	0.10	0.11	0.11	0.11	0.11	0.11	0.12
A3	0.11	0.11	0.04	0.04	0.04	0.04	0.04	0.04	0.04
A4	0.04	0.05	0.04	0.14	0.15	0.16	0.14	0.18	0.13
A5	0.12	0.14	0.17	0.14	0.10	0.10	0.09	0.11	0.09
A6	0.09	0.10	0.12	0.09	0.10	0.09	0.09	0.09	0.08
A7	0.08	0.09	0.10	0.08	0.09	0.09	0.11	0.13	0.10
A8	0.10	0.11	0.13	0.11	0.12	0.11	0.11	0.06	0.05
A9	0.05	0.06	0.07	0.05	0.05	0.06	0.05	0.06	0.17
Weight	0.23	0.12	0.06	0.11	0.12	0.09	0.1	0.1	0.07
Indicator	A1	A2	A3	A4	A5	A6	A7	A8	A9

Source: Created by the authors.
Note: The priority vector for the governance axis is presented below the matrix.

the selection of six out of the nine proposed indicators, with priority weights of 0.23, 0.12, 0.11, and 0.10, among others (Table 8.4, values in bold).

8.4.2.1 Assessment Model for ICT Indicators

Drawing on the analysis of the results obtained in this study; the AMII is proposed as a self-evaluation tool for assessing the availability and use of ICTs in tourist destinations seeking STD certification. The structure of the AMII is illustrated in Figure 8.4 and

TABLE 8.4
Final Weights Assigned to Governance Indicators According to AHP Analysis

Governance Indicators	Weight
Implementation of a strategic master plan for smart tourism destinations	**0.23**
Existence of a system to know the degree of tourist satisfaction in the destination	**0.12**
Elaboration and diffusion of a periodic report of activities through different channels	0.06
Percentage of city services online. Exclusive or integrated platform in the municipal web	**0.11**
Existence of platforms for information exchange and public-private partnerships	**0.12**
Existence of an office with its own staff of the City Council as a coordinating entity for actions to set up an STD (smart tourism office)	0.09
Number of workers trained in smart tourism destinations per year for the development of new management skills related to STD or ICT	**0.10**
Percentage of workforce working in ICT	**0.10**
Number of tourist events and economic activities scheduled per year of matters of common interest on official website of the destination	0.07

Source: Created by the authors.

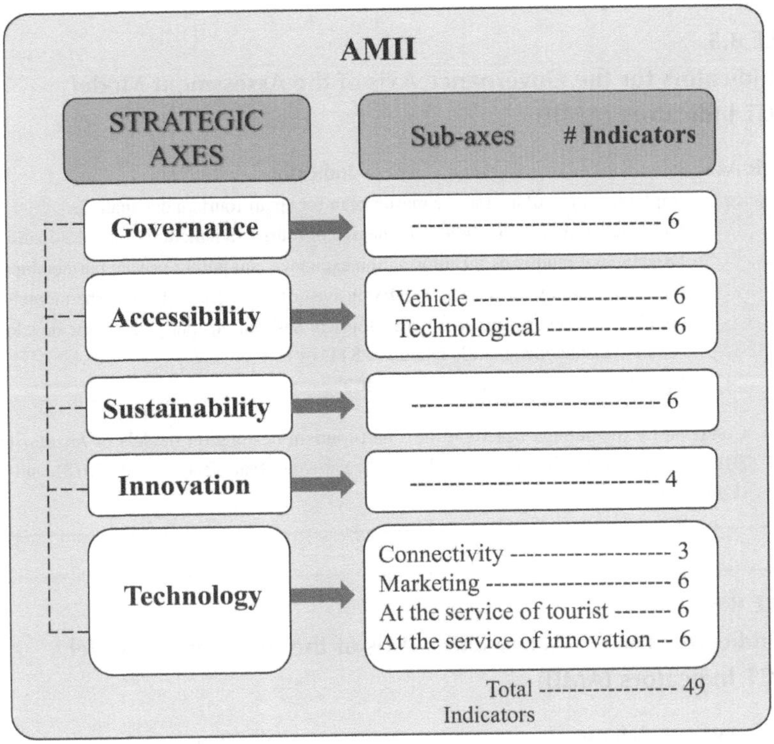

FIGURE 8.4 Structural configuration of the assessment model for ICT indicators (AMII), with strategic axes and corresponding ICT indicators.

Source: Developed by the authors.

comprises five strategic axes: Governance, Accessibility, Sustainability, Innovation, and Technology. Due to thematic overlaps between indicators associated with the Accessibility and Technology axes, sub-axes were introduced to group indicators more coherently. Figure 8.4 also presents the number of ICT indicators associated with each strategic axis.

To assess the governance dimension of a prospective STD, the indicators listed in Table 8.5 should be applied. These indicators address key aspects of ICT planning and management within the destination and reflect the involvement of multiple tourism stakeholders. They emphasize the importance of information exchange platforms, ICT-based services, and mechanisms for public–private collaboration.

The Accessibility axis comprises two sub-axes: Vehicle Accessibility and Technological Accessibility, each containing six indicators (Table 8.6). The Vehicle Accessibility sub-axis addresses aspects such as parking availability, informational signage, public and private transport services, and other modes of local mobility. The Technological Accessibility sub-axis includes indicators related to broadband access, digital promotion efforts, compliance with accessibility regulations, and the provision of virtual tourist information.

TABLE 8.5
ICT Indicators for the Governance Axis of the Assessment Model for ICT Indicators (AMII)

Strategic Axis	Indicators
Governance	Implementation of a strategic master plan for smart tourism destinations
	Existence of a system to know the degree of tourist satisfaction in the destination
	Existence of platforms for information exchange and public-private partnerships
	Percentage of city services online. Exclusive or integrated platform in the municipal web
	Number of workers trained in smart tourism destinations per year for the development of new management skills related to STD or ICT
	Percentage of workforce working in ICT

Source: Compiled by the authors based on the smart tourism destination models of *Invattur* (Invattur, 2015), *Competitive* (Cimbaljević et al., 2018), *Systemic* (Ivars-Baidal et al., 2018), and *Synergic* (Cabral, 2018).

TABLE 8.6
ICT Indicators for the Accessibility Axis of the Assessment Model for ICT Indicators (AMII)

Strategic Axis	Sub-axes	Indicators
Accessibility	Vehicle	Percentage of public transport monitored in real time and accessible to the public.
		Existence of a vehicle tracking system that provides real-time information on the transport network and can be distributed to end-user devices.
		Percentage of public transport covered by a single payment system.
		Number of bicycles available per 100,000 inhabitants.
		Percentage of car parks equipped with electronic payment.
		Percentage of autonomous vehicles in the vehicle fleet.
	Technological	Percentage of target area with sufficient access to fixed broadband.
		Existence of digital promotion of tourist attractions.
		Certification of the WAI protocol, standard UNE 139803:2004 and others (web accessibility compliance).
		Percentage of the area of the city with sectors covered by the telecommunications network.
		Availability of: audio guides, sign guides, accessibility guides, guides in braille, adapted tourist visits, others.
		Availability of updated virtual tourist information adapted for people with disabilities. UNE 170001-2 Accessibility and others. Exterior and interior accessibility. Counters, showcases, totems, and adapted furniture. Trained staff.

Source: Compiled by the authors based on the smart tourism destination models of *Invattur* (Invattur, 2015), *Competitive* (Cimbaljević et al., 2018), *Systemic* (Ivars-Baidal et al., 2018), and *Synergic* (Cabral, 2018).

TABLE 8.7
ICT Indicators for the Sustainability Axis of the Assessment Model for ICT Indicators (AMII)

Strategic Axis	Indicators
Sustainability	Percentage of public buildings with smart energy systems and indoor air quality monitoring equipment. Applications of inmotic to public buildings.
	Percentage of drinking and environmental water with real-time quality monitoring.
	Percentage of containers with telemetry. Reduction of service delivery costs and improvement of environmental parameters.
	Monitoring system connected with the urban approach and Agenda 21. Monitoring and alerts air and water quality.
	Percentage of the distribution and wastewater network monitored with smart systems.
	Percentage of coverage in the city of real-time online traffic alerts. Measures to reduce energy consumption and air and noise pollution, as well as the cost of traffic congestion.

Source: Compiled by the authors based on the smart tourism destination models of *Invattur* (Invattur, 2015), *Competitive* (Cimbaljević et al., 2018), *Systemic* (Ivars-Baidal et al., 2018), and *Synergic* (Cabral, 2018).

For the AMII Sustainability axis, the indicators described in Table 8.7 are proposed for use. These indicators primarily assess the level of integration of intelligent systems in applications such as energy management in public buildings, the monitoring of water quality and the sewage network, as well as the use of telemetry-enabled waste containers and online urban traffic alert systems.

The Innovation axis evaluates the presence of advanced digital solutions implemented within the tourist destination. These solutions may include technologies related to tourism offerings, business management systems, hardware and software infrastructure, and electronic devices used in tourist facilities for service access and payment. The full set of indicators for this axis is presented in Table 8.8.

Finally, the Technology axis of the AMII comprises four sub-axes: (1) Connectivity, (2) Marketing, (3) Technology for the Tourist, and (4) Technology for Innovation (Table 8.9).

- **Connectivity** includes initiatives by public and private entities to provide free Wi-Fi access, as well as the development of applications and other technological solutions for tourists.
- **Marketing** evaluates the presence of action plans, online promotional strategies, and tools for analyzing tourist behavior.
- **Technology for the Tourist** focuses on user-oriented digital, including engaging and responsive websites, as well as co-creation tools such as cloud-based tour guides, touchscreen displays, and virtual information kiosks in destination offices.
- **Technology for Innovation** assesses ongoing website updates, digital analytics, information automation in tourism offices, the availability of Open Data, and the implementation of sensor systems throughout the destination.

TABLE 8.8
ICT Indicators for the Innovation Axis of the Assessment Model for ICT Indicators (AMII)

Strategic Axis	Indicators
Innovation	Existence of innovative digital solutions to improve the tourist offer.
	Existence of an innovative strategy for managing the business-tourist relationship.
	Existence of hardware and software that supports internet of things and cloud services (apps, tablets, touch screen, etc.).
	Availability of electronic devices to facilitate service access and payment in tourist facilities.

Source: Compiled by the authors based on the smart tourism destination models of *Invattur* (Invattur, 2015), *Competitive* (Cimbaljević et al., 2018), *Systemic* (Ivars-Baidal et al., 2018), and *Synergic* (Cabral, 2018).

TABLE 8.9
ICT Indicators of the Strategic Axis Technology of the Assessment Model for ICT Indicators (AMII)

Strategic Axis	Sub-axes	Indicators
Technology	Connectivity	Number of public initiative free Wi-Fi access points.
		Number of private initiative free Wi-Fi access points.
		Existence of applications, technological solutions, etc. for tourists. (tools or services that allow Wi-Fi accessibility in destination).
	Marketing	Designing innovative campaigns, sending videos, animations, etc. (Inspire).
		Existence of a marketing plan, analysis of tourist behavior and online tourism promotion (portal, social networks, multimedia, apps with destination information).
		Existence of promotional merchandising material for the tourist. Preferably the payment method will be based on new technologies to boost/streamline this sales process.
		Existence of web section with downloads of informational/promotional material. Existence of technological solutions for the download. Destination app.
		Existence of newsletters, promotions, last-minute offers, new products, etc. (Inspire, loyalty).
		Search engine optimization (SEO), search engine marketing (SEM), blogs, etc. Social bookmarking. Participate in discussions, blog posts, etc.
	For the Tourist	Existence of an attractive, simple and fast website. Adapted to all devices. Promote processes of interaction and co-creation. With possibility to make reservations. (web tourist).
		The existence of NFC tags and QR codes to access information about nearby points of interest via mobile devices.
		Presence in social networks with major importance for tourism promotion.

(Continued)

TABLE 8.9 ***(Continued)***
ICT Indicators of the Strategic Axis Technology of the Assessment Model for ICT Indicators (AMII)

Strategic Axis	Sub-axes	Indicators
		Existence of guidebooks in the cloud (audio and/or video guides).
		Existence of touch screens and virtual information kiosks in information office and focal points of the destination. (24 hours information channels).
		Existence of interactive scorecard with key indicators for consultation and use of destination data.
	For Innovation	Existence of mechanisms for the permanent updating of digital information. Degree of updating of information.
		Online reputation monitoring. Google Analytics, Google Alerts, Tweel Deck, Social Mention, etc. Go beyond the use of social networks: analyze what is said about the destination and what is the profile/ motivations of our digital visitor.
		Automation of tourist information. Collection of data from tourist offices: implementation of a system to automate the collection of tourist information at tourism offices (requests for information by type, number of users, etc.). Data analysis of tourist offices: analysis of information obtained for decision-making in the destination.
		Existence of a social media plan with the collaboration of private initiative.
		Existence of open data. Intention to open data: the publication of open data would be valued as a greater willingness to open tourism data. Open data publication: open data is published. Published data quality level: data quality level according to 5-star re-use scheme.
		Sensors of various types, such as traffic management, which have applications to tourism management. Use of visitor geolocation (GPS).

Source: Compiled by the authors based on the smart tourism destination models of *Invattur* (Invattur, 2015), *Competitive* (Cimbaljević et al., 2018), *Systemic* (Ivars-Baidal et al., 2018), and *Synergic* (Cabral, 2018).

8.5 CONCLUSIONS

STDs are tourism-oriented locations that must meet specific criteria to obtain STD certification, as outlined by various established conceptual models. Within these models, the "smart" component is closely linked to ICTs. To obtain this certification, a tourist destination must undergo an evaluation process that includes a set of indicators, which may or may not be directly related to ICT use. This technological component plays a critical role in the diagnostic assessment of destinations seeking STD recognition, as it is integrated across all strategic axes of the models.

In response to the need for consistent evaluation criteria, this research conducted a comparative analysis of well-defined and published STD conceptual models. Based on this analysis, a self-assessment proposal was developed for tourist destinations pursuing STD certification. The observed similarities among the models underscore the considerable presence and cross-cutting nature of ICT-related indicators across strategic axes. This led to the formulation of a practical tool, the AMII, designed to facilitate the structured evaluation of ICT implementation in tourism contexts.

The assessment model developed in this research is derived from existing STD conceptual models and focuses exclusively on ICT-related indicators. The AMII is structured around five strategic axes, encompassing a total of 49 indicators distributed as follows: Governance (6 indicators), Accessibility (12), Sustainability (6), Innovation (4), and Technology (21). The strength of the AMII lies in its practical applicability as a self-assessment tool that supports the formulation of innovative tourism management strategies and facilitates progress toward STD certification.

This research introduces an innovative methodological approach by integrating the AHP with expert panel consultation. This combination enabled the objective and well-founded selection of ICT indicators for evaluating STDs, ensuring methodological rigor and transparency in the indicator development process.

The integration of ICTs is increasingly essential for the future development and global competitiveness of tourist destinations. Destinations that fail to adopt these tools risk falling behind in delivering quality visitor experiences and in achieving efficient tourism management. In this context, the AMII offers a practical framework for assessing the availability and use of ICT in tourist destinations. Its application can support both strategic decision-making and the design of public policies aimed at fostering innovation and sustainability in tourism governance.

REFERENCES

Alsharif, A., Isa, S. M. & Alqudah, M. N. (2024). Smart tourism, hospitality, and destination: A systematic review and future directions. Journal of Tourism and Services, 15(29), 72–110. https://doi.org/10.29036/jots.v15i29.746

Berges-García, A., Meneses-Chaus, J. M. & Martínez-Ortega, J. F. (2016). Methodology for evaluating functions and products of technological surveillance and competitive intelligence (VT/IC) and its implementation through the web. The Professional of Information, 25(1), 103. https://doi.org/10.3145/epi.2016.ene.10

Cabral Ghizoni De Sousa, T. (2018). Model of Competitiveness of Smart Tourism Destinations [Doctoral Thesis. University of Alicante]. Institutional Repository of the University of Alicante. http://hdl.handle.net/10045/77603

Carballido, A. & Guevara-Plaza, A. (2021). El concepto de Smart Destination en la investigación turística. Revisión sistemática de la literatura para su definición y normalización. Cuadernos de Turismo, 48, 301–323. https://doi.org/10.6018/turismo.492781

Cerdá-Mansilla, E., Tussyadiah, I., Campo, S. & Rubio, N. (2024). Smart destinations: A holistic view from researchers and managers to tourists and locals. Tourism Management Perspectives, 51, 1–16. https://doi.org/10.1016/j.tmp.2024.101223

Cimbaljević, M., Stankov, U. & Pavluković, V. (2018). Going beyond traditional destination competitiveness–reflections on a smart destination in current research. Current Issues in Tourism, 22(20), 2472–2477. https://doi.org/10.1080/13683500.2018.1529149

EU (2019). Compendium of best practices. 2019 European Capital of Smart Tourism competition.: Vol. I.

Gretzel, U., Sigala, M., Xiang, Z. & Koo, C. (2015). Smart tourism: Foundations and developments. Electronic Markets, 25(3), 179–188. https://doi.org/10.1007/s12525-015-0196-8

Hernández-Sampieri, R., Fernández, C. & Baptista, P. (2014). Research Methodology (6th ed.). McGraw-Hill.

Herrera-Prado, A. L., (2021). Impact of Information and Communication Technologies on the tourism activity of Mazatlán, Sinaloa: a focus on Smart Tourism Destination [Doctoral thesis]. Institutional Repository of the Universidad Autónoma de Occidente. https://uadeo.mx//wp-content/uploads/2022/06/2021-12-Tesis-ALHP.pdf

Herrera-Prado, A. L., Velarde-Valdez, M. & Olmos-Martínez, E. (2024a). Participation of ICT in the models of Smart Tourism Destinations. Tourist Research (27), pp. 1–28. https://doi.org/10.14198/INTURI.20111

Herrera-Prado, A. L., Zazueta-Hernández, M. G. & León-Castro, E. (2024b). Effects of the implementation of the Smart Tourism Destination model in the magical town of Tequila, Jalisco, Mexico. In Studies in systems, decision and control, pp. 1–19. https://doi.org/10.1007/978-3-031-69382-3_1

Iglesias, T., García, B. & Fernández, A. (2018). Smart tourism destinations within urban tourism: The case of León. Tourism Studies, 216(2o S 2018), 107–132. https://dialnet.unirioja.es/servlet/catart?codigo=7042606

Invattur (2015). Smart Tourism Destinations. Operational Manual for the Configuration of Smart Destinations. Valencian Tourism Agency. https://invattur.es/uploads/entorno_37/ficheros/62690a0d5bc1b967049880.pdf

Ivars-Baidal, J., Celdrán Bemabeu, M. & Femenia-Serra, F. (2017). Guide for the Implementation of Smart Tourism Destinations - Comunidad Valenciana. University Institute of Tourist Research (IUIT) University of Alicante. https://www.researchgate.net/publication/323880784_Guia_de_Implantacion_de_Destinos_Turisticos_Inteligentes-Comunitat_Valenciana

Ivars-Baidal, J. A., Celdrán-Bernabeu, M. A., Femenia-Serra, F., Perles-Ribes, J. F. & Giner Sánchez, D. (2021). Measuring the progress of smart destinations: The use of indicators as a management tool. Journal of Destination Marketing and Management, 19(November 2020). https://doi.org/10.1016/j.jdmm.2020.100531

Ivars-Baidal, J. A., Celdrán-Bernabeu, M. A., Mazón, J. N. & Perles-Ivars, ÁF. (2019). Smart destinations and the evolution of ICTs: A new scenario for destination management? Current Issues in Tourism, 22(13), 42. https://doi.org/10.1080/13683500.2017.1388771

Ivars-Baidal, J. A., Navarro Jurado, E., Celdrán Bernabeu, M., Perles Ribes, J. & Perea Medina, M. (2018). The comprehensive approach of Smart Tourism Destinations (STD) in coastal areas: Scope, Progress, and Limitations. IV Congress of Smart Cities, May 6.

Ivars-Baidal, J. A., Solsona Monzonís, F. J. & Giner Sánchez, D. (2016). Tourism management and information and communication technologies (ICT): The new approach of smart destinations. Documents d'Anàlisi Geogràfica, 62(2), 327–346. https://doi.org/10.5565/rev/dag.285

Kontogianni, A., Kabassi, K. & Alepis, E. (2018). Designing a Smart Tourism Mobile Application: User Modelling Through Social Networks' User Implicit Data. In S. Staab (Ed.), SocInfo 2018, LNCS (pp. 148–158). Springer International Publishing. https://doi.org/10.1007/978-3-030-01159-8

López de Ávila, A., Lancis, E., García, S., Alcantud, A., García, B. & Muñoz, N. (2015). Report on Smart Tourism Destinations: Building the Future. SEGITTUR. http://www.thinktur.org/media/Libro-Blanco-Destinos-Turisticos-inteligentes-construyendo-el-futuro.pdf

Moliner-Tena, M. A., Monteserín-Abella, O., Callarisa-Fiol, L. J., & Ferreres Bonfill, J. B. (2026). Smart tourist destination: A literature review and future agenda. Journal of Hospitality and Tourism Insights, 9(3), 1065–1085. https://doi.org/10.1108/JHTI-04-2025-0517

Montecé-Macías, M. & Marcial-Chang, D. (2023). Environmental tourism sustainability strategies for positioning Guayaquil as a smart tourism destination. International Journal of Management, Innovation and Sustainable Tourism, 3(1), 12–24. https://doi.org/10.51260/rigistur.v3i1.386

Saaty, R. W. (1987). The analytic hierarchy process - what it is and how it is used. Mathematical Modelling, 9(3–5), 161–176. https://doi.org/10.1016/0270-0255(87)90473-8

SEGITTUR (2018). The Spanish Strategy for Smart Tourism Destinations. European Conferences. Sustainability and Smart Territories, 40. https://retodemografico.dip-caceres.es/wp-content/uploads/2022/06/03_Estrategia_Destinos_Tcos_Inteligentes.pdf

SEGITTUR (2022). Los 5 ejes que definen los destinos turísticos Inteligentes. Recuperado de. https://www.destinosinteligentes.es/los-5-ejes-que-definen-los-destinos-turisticos-inteligentes/

SEGITTUR (2025) Smart tourism destinations. Retrieved June 2025 from https://www.destinosinteligentes.es/que-es-dti/

UN World Tourism Organization (UNWTO) (2025). Digital transformation. (accessed Mayo 2025). https://www.unwto.org/digital-transformation

Wang, X. (2024). Construction of smart tourism system integrating tourist needs and scene characteristics. Systems and Soft Computing, 6, 200168. https://doi.org/10.1016/j.sasc.2024.200168

Wang, X., Li, X. R., Zhen, F. & Zhang, J. H. (2016). How smart is your tourist attraction?: Measuring tourist preferences of smart tourism attractions via a FCEM-AHP and IPA approach. Tourism Management, 54(November 2017), 309–320. https://doi.org/10.1016/j.tourman.2015.12.003

ANNEXES

ANNEX 8.1

Strategic Axes and ICT Indicators in the *Invattur* Model

Strategic Axes	ICT Indicators
Accessibility	Information service adapted for people with disabilities.
	Web accessibility compliance (WAI protocol).
	Promotion of accessible tourism.
Connectivity and sensorization	Network quality connectivity (bandwidth).
	Free Wi-Fi with guaranteed QoS in public places (QoS = quality of service).
	Free Wi-Fi with guaranteed QoS in companies in the tourism sector (hotels, restaurants, shops, etc.).
	Sensorization applied to tourism management.
Governance	Implementation of an STD master plan.
	Smart tourism office.
	Dissemination of periodic activity reports.
	Existence of public-private partnership platforms.
	E-Government.
	Quality perceived by tourists.
	Training in the development of new management skills related to smart tourism destinations or ICT.
Tourist information	Promotional material in digital format.
	Information channels 24 hours.
	Virtual tourist information.
	Tourist website.
	Social media.

(Continued)

ANNEX 8.1 *(Continued)*
Strategic Axes and ICT Indicators in the *Invattur* Model

Strategic Axes	ICT Indicators
	Inspirational videos.
	Tools or services that allow Wi-Fi accessibility in the destination.
	Digital information technologies (e.g., QR codes, NFC, RFID).
	Technological tools for the tourist.
	Merchandizing.
	Stock of promotional material.
Innovation	Incorporation of ICT (online consultation and procedures).
	Collaborative innovation processes.
Online marketing	Brand and media monitoring.
	Social media plan.
	Positioning and SEO actions
	Online advertising.
	CRM strategies and email marketing.
	Viral marketing.
Information system/smart tourism	Automation of information in tourist offices (ATO).
	Characterization, behavior and satisfaction (CBS) of tourism demand.
	Business barometer (BB).
	Web analytics and social networks (WSN).
	Smart destination platform for information integration.
	Mechanisms for the permanent updating of digital information.
	Open data (OD).
	Diffusion and visualization of data.
Territorial sustainability	Urban mobility.
	Energy efficiency.
	Waste collection and treatment.
	Water supply, purification, and reuse.
	Management of public buildings.
	Existence of a system of sustainability indicators.

Source: Compiled by the authors based on Invattur (2015).

ANNEX 8.2
Strategic Axes and ICT Indicators in the *Competitive* Model

Strategic Axes	ICT Indicators
Competitiveness	Development of e-government/open administration.
	Quality network connectivity (bandwidth).
	Quality network coverage (fixed broadband internet).
	Training in the development of new management skills related to smart tourism destinations or ICT.
	Incorporation of ICT (online consultation and procedures).

(Continued)

ANNEX 8.2 *(Continued)*
Strategic Axes and ICT Indicators in the *Competitive* Model

Strategic Axes	ICT Indicators
Tourism components	Tourism telecommunications services.
	Digital promotion of tourism activities.
	Booking and purchasing tour packages online.
	Electronic devices to facilitate access and payments in tourist facilities.
	Digital promotion of tourist attractions.
	Digital transport services.
	Destination management systems (DMS).
	Computer systems to measure customer satisfaction.
	Augmented reality (AR) allows visitors to experience digital recreation of tourist sites and time travel.
	The vehicle tracking system provides real-time information on the transport network and could be distributed to end-user devices.
	The hotel must be able to predict energy demand for construction and perform energy audits based on its environmental management.
	A multilingual application that provides a range of services, such as an electronic travel guide, which also offers a number of packages available to tourists.
	NFC tags and QR codes to access information about nearby points of interest via mobile devices.
	Tourists can register their complaints through a claims management system that is supported by various ICT channels, such as SMS or mobile applications, which could route them directly to the appropriate officials.
Smart	Development of e-government/open administration.
	Training in the development of new management skills related to STD or ICT.
	Urban mobility (online alerts).
	Urban mobility (real-time monitoring of public transport).
	Urban mobility (bike sharing).
	Urban mobility (single payment for public transport).
	Urban mobility (parking with electronic payment).
	Urban mobility (parking with real-time availability).
	Urban mobility (smart traffic lights).
	Urban mobility (interactive map in real time).
	Urban mobility (autonomous vehicles).
	Urban mobility (connection of public transport for management and monitoring).
	Management of public buildings (intelligent energy).
	Management of public buildings (air monitoring).
	Energy efficiency.
	Waste collection and treatment (containers).
	Waste collection and treatment (collection at home).
	Water supply, purification, and reuse (monitoring of drinking quality).
	Water supply, purification, and reuse (environmental quality monitoring).
	Water supply, purification, and reuse (distribution monitoring).
	Water supply, purification, and reuse (smart buildings).
	Water supply, purification, and reuse (houses).
	Water supply, purification, and reuse (monitoring of wastewater lines).

(Continued)

ANNEX 8.2 *(Continued)*
Strategic Axes and ICT Indicators in the *Competitive* Model

Strategic Axes	ICT Indicators
	Existence of a system of sustainability indicators (air quality).
	Information service adapted to people with disabilities.
	Quality network connectivity (bandwidth).
	Quality network connectivity (telecommunication).
	Wi-Fi-free with QoS guaranteed in public places.
	Open Data (contracts).
	Incorporation of ICT (labor force).
	Incorporation of ICT (online medical record).
	Incorporation of ICT (medical appointments).
	Existence of a system of sustainability indicators (air and water quality alerts).
	Incorporation of ICT (online consultation and procedures).
	Open data (consultations).
	E-government.
	E-government (response time).
	E-government (falling time).
	Promotion of accessible ICT.
	Incorporation of ICT (recreation).
	Sensorization applied to safety.
	Incorporation of ICT (culture).
	Incorporation of ICT (digitalized culture).
	Incorporation of ICT (e-book).
	Urban mobility (food suppliers map).
	E-government (building permits).
Digital services	Cloud services.
	Internet of things.
	Internet service system for end users.

Source: Compiled by the authors based on Cimbaljević et al. (2018).

ANNEX 8.3
Strategic Axes and ICT Indicators in the *Systemic* Model

Strategic Axes	ICT Indicators
Accessibility	Information service adapted for people with disabilities.
	Compliance with tourist web accessibility (WAI protocol).
	Promotion of accessible tourism.
Connectivity and sensorization	Quality connectivity to the network of tourism enterprises.
	Public Wi-Fi with QoS in places of tourist influx.
	Sensorization applied to tourism management.

(Continued)

ANNEX 8.3 *(Continued)*
Strategic Axes and ICT Indicators in the *Systemic* Model

Strategic Axes	ICT Indicators
Governance	Existence of interdepartmental coordination procedures in local government.
	Implementation of an STD master plan.
	Design and implementation of annual operational plans for tourism management.
	Monitoring the degree of compliance with operational plans.
	Regular dissemination of tourism management reports.
	Existence of public-private partnership structures.
	Degree of development of e-government.
	Implementation of a destination-based quality management system.
	Development of social awareness campaigns around tourism.
Tourist information	Promotional material in digital format.
	Virtual information channels 24 hours.
	Quality official tourist website.
	Social media presence.
	Quality certifications in tourist offices.
	Technologies linking tourist spaces/resources with digital information (QR, NFC, RFID, etc.).
	Electronic devices offered to tourists.
Innovation	Innovative municipal tourist organization (development of innovation projects).
	Measures to promote innovation and tourism entrepreneurship.
	Tourism enterprises with implemented innovation management systems.
	Degree of incorporation of ICT in tourist enterprises in the municipality.
Online marketing	Monitoring the online reputation of the destination.
	Existence of a social media plan.
	Positioning and SEO actions.
	Online advertising
	Relationship marketing strategy (customer relationship marketing)
Information system	Automation of data collection in tourist information offices.
	Demand characterization, behavior, and satisfaction studies.
	Business confidence indicators.
	Tourism web analytics and social media.
	Integrated smart information platform.
	Opening of municipal data (open data).
	Data diffusion and visualization tools (such as Scorecards).
	Spatial data infrastructure (SDI) adapted to the INSPIRE directive.
Sustainability	Measures to promote sustainable mobility.
	Energy efficiency.
	Collection and treatment of waste.
	Water supply, purification, and reuse.
	Noise pollution.
	Existence of a system of sustainability indicators.
	Raising awareness of sustainability for residents.
	Raising awareness of sustainability for tourists.

Source: Compiled by the authors based on Ivars-Baidal et al. (2018).

ANNEX 8.4
Strategic Axes and ICT Indicators in the *Synergic* Model

Strategic Axes	ICT Indicators
Tourist attractions and supply conditions	Number of annual events on topics of common interest listed on the destination's official website.
Competitive and social conditioning of the urban environment	Quality network connectivity (bandwidth).
	Urban mobility (real-time monitoring of public transport).
Demand and economic impact	Characterization, behavior, and satisfaction of tourist demand (CBS).
Governance and strategic management	Implementation of an STD master plan.
	Existence of a smart tourism office.
	Web analytics and social networks (WSN).
	Training in the development of new management skills related to STD or ICT.
	Development of electronic administration/open administration.
	Incorporation of ICT (consultation and online procedures).
Innovation	Collaborative innovation processes.
	Economic activities linked to innovation.
Sustainability and environment	Existence of a system of sustainability indicators (air quality).
	Existence of a system of sustainability indicators.
ICT, marketing, and information system	Free Wi-Fi with guaranteed QoS in public places.
	Free Wi-Fi with guaranteed QoS in public places (speed).
	Internet service system for end users.
	Virtual tourist information.
	Web analytics and social media (WSM).
	Characterization, behavior, and satisfaction of tourist demand (CBS).
	Business-tourist relationship management strategy.
	Open data (OD).
	Smart destination information integration platform.
	Social media plan.
	Sensorization applied to tourism management (GPS).
	Sensorization applied to tourism management.
	Brand and media monitoring.
Accessible tourism	Information service adapted to people with disabilities.

Source: Compiled by the authors based on Cabral (2018).

9 Modeling Uncertainty in Decision Systems Using Neutrosophic Logic and Blockchain-Based Traceability

Ruhit Bardhan, Dharmik Chauhan, Manoj Sahni, and Ajoy Kanti Das

9.1 INTRODUCTION

Decision-making is a fundamental process in various domains including business, healthcare, engineering, and governance. However, real-world decision scenarios often involve uncertainty arising from incomplete information, vague data, conflicting evidence, and subjective judgments. Traditional decision-making frameworks based on classical logic or probability theory may not adequately capture all aspects of uncertainty, particularly when dealing with indeterminate or paradoxical information. Blockchain technology has emerged as a revolutionary paradigm for establishing trust and transparency in distributed systems without requiring central authority. Its key features including immutability, decentralization, consensus mechanisms, and smart contracts make it an ideal platform for implementing decision systems that require auditability and trustworthiness.

Neutrosophic logic, introduced by Smarandache (Smarandache, 1999; Smarandache and Jdid, 2023; Fujita and Smarandache, 2025), generalizes classical logic, fuzzy logic, and intuitionistic fuzzy logic by explicitly representing three components: truth $(\mathcal{T})$, indeterminacy $(\mathcal{I})$, and falsity $(\mathcal{F})$. This tripartite structure enables more nuanced modeling of uncertainty, particularly in situations where information is incomplete, contradictory, or indeterminate. The integration of blockchain technology with neutrosophic logic is motivated by several fundamental limitations of existing decision systems. Conventional decision models are inadequate in handling indeterminate and conflicting information. Centralized decision-making frameworks often suffer from transparency and weak auditability. Distributed consensus mechanisms require reliable methods for representing and managing uncertainty. Moreover, critical application domains demand both high security and rigorous uncertainty quantification. The proposed integration addresses these challenges by combining transparent, tamper-resistant data management with a flexible uncertainty modeling framework.

 DOI: 10.1201/9781003724902-10

Although several studies have explored fuzzy and intuitionistic fuzzy decision models within blockchain frameworks (Cali et al., 2022; Nabeeh and Tantawy, 2023), most existing approaches fail to account for the independent nature of indeterminacy. Fuzzy systems handle ambiguity, but not paradoxical or conflicting evidence that often arises in decentralized decision environments. Intuitionistic fuzzy sets introduce hesitation, yet still merge indeterminacy with uncertainty, which limits the modeling precision in distributed consensus. This research bridges gap by proposing a hybrid Neutrosophic-Blockchain Framework (Cali et al., 2022; Vafadarnikjoo et al., 2023; Aliahmadi and Nozari, 2023), where neutrosophic components $(\mathcal{T}, \mathcal{I}, \mathcal{F})$ explicitly quantify truth, indeterminacy, and falsity during block validation. The framework not only increases transparency and reliability but also introduces a mathematical foundation for handling indeterminate states within consensus protocols.

This research introduces decision-making architecture that combines blockchain infrastructure with neutrosophic logic. The proposed model addresses uncertainty, partial knowledge, and inconsistency inherent in decentralized systems. A customized consensus strategy based on neutrosophic principles is formulated for blockchain environments. The consensus process evaluates decisions using truth, indeterminacy, and falsity components to enhance validation reliability. Smart contract mechanisms are implemented to support the automatic execution of neutrosophic decision outcomes. These mechanisms promote transparency, data integrity, and decentralized governance within the blockchain network. Multiple application-oriented case studies are conducted to illustrate the practical relevance of the proposed approach. The findings demonstrate that neutrosophic reasoning effectively handles complex decision scenarios in real-world blockchain applications. A systematic comparison is performed with traditional and fuzzy decision-making techniques. The results indicate that the proposed framework offers greater adaptability and robustness when dealing with uncertain and ambiguous information.

In this chapter, Section 9.2 describes the preliminaries of the neutrosophic sets and fundamentals of blockchain. Section 9.3 includes the main results of this study. Section 9.4 shows real-life applications of the blockchain using the neutrosophic sets system. Section 9.5 concludes the study and highlights potential directions for future research.

9.2 PRELIMINARIES

This section provides a rigorous overview of the fundamental definitions and essential principles underlying neutrosophic theory and blockchain systems.

9.2.1 Neutrosophic Set

Neutrosophic set theory, introduced by Smarandache, extends traditional uncertainty modeling by representing each element with three independent components: truth, indeterminacy, and falsity. Unlike classical and fuzzy sets that rely on a single membership value, neutrosophic sets allow all three components to coexist, providing a more realistic way to express incomplete, inconsistent, or vague information, especially in decision-making applications.

Definition 1 (Neutrosophic set)

Let $\mathcal{U}$ be a universe of discourse. A neutrosophic set (Smarandache, 1999; Wang et al., 2010; Ye, 2014; Parikh and Sahni, 2020; Bardhan et al., 2025) $\mathcal{A}$ in $\mathcal{U}$ is characterized by three membership functions: truth membership $\mathcal{T}_{\mathcal{A}}(x)$, indeterminacy membership $\mathcal{I}_{\mathcal{A}}(x)$ and falsity membership $\mathcal{F}_{\mathcal{A}}(x)$, where, $x \in \mathcal{U}$ and $\mathcal{T}_{\mathcal{A}}, \mathcal{I}_{\mathcal{A}}, \mathcal{F}_{\mathcal{A}} : \mathcal{U} \to [0,1]$. The condition $0 \le \mathcal{T}_{\mathcal{A}}(x) + \mathcal{I}_{\mathcal{A}}(x) + \mathcal{F}_{\mathcal{A}}(x) \le 3$ holds for all $x \in \mathcal{U}$.

Definition 2 (Single-valued neutrosophic number (SVNN))

A single-valued neutrosophic number (SVNN) (Smarandache, 1999; Wang et al., 2010; Parikh et al., 2022) is denoted as $\mathcal{N} = \langle \mathcal{T}, \mathcal{I}, \mathcal{F} \rangle$, where, $\mathcal{T}, \mathcal{I}, \mathcal{F} \in [0,1]$ and $0 \le \mathcal{T} + \mathcal{I} + \mathcal{F} \le 3$.

Neutrosophic Operations

Using **Definition 1** of neutrosophic set, for two SVNNs $\mathcal{N}_1 = \langle \mathcal{T}_1, \mathcal{I}_1, \mathcal{F}_1 \rangle$ and $\mathcal{N}_2 = \langle \mathcal{T}_2, \mathcal{I}_2, \mathcal{F}_2 \rangle$, a neutrosophic operation (Smarandache, 1999) is defined as,

$$\mathcal{N}_1 \oplus \mathcal{N}_2 = \langle \mathcal{T}_1 + \mathcal{T}_2 - \mathcal{T}_1\mathcal{T}_2, \mathcal{I}_1\mathcal{I}_2, \mathcal{F}_1\mathcal{F}_2 \rangle \tag{9.1}$$

$$\mathcal{N}_1 \otimes \mathcal{N}_2 = \langle \mathcal{T}_1\mathcal{T}_2, \mathcal{I}_1 + \mathcal{I}_2 - \mathcal{I}_1\mathcal{I}_2, \mathcal{F}_1 + \mathcal{F}_2 - \mathcal{F}_1\mathcal{F}_2 \rangle \tag{9.2}$$

$$\lambda\mathcal{N} = \langle 1 - (1-\mathcal{T})^{\lambda}, \mathcal{I}^{\lambda}, \mathcal{F}^{\lambda} \rangle, \ \lambda > 0 \tag{9.3}$$

9.2.2 Blockchain Fundamentals

A blockchain is a distributed digital record, where information is organized into a sequence of blocks connected through cryptographic hash values. Each block includes metadata in the header along with a collection of validated transactions. Since the ledger is shared across multiple nodes, a consensus mechanism is used to make sure that every participant reaches the same view of the current blockchain state, even without a central authority.

Definition 3 (Blockchain)

A blockchain (Nakamoto, 2008) is a decentralized and distributed ledger composed of an ordered sequence of blocks.

$$\mathcal{B}_i = \{\mathcal{B}_0, \mathcal{B}_1, \ldots, \mathcal{B}_n\},$$

where each block $\mathcal{B}_i$ is cryptographically linked to its predecessor. Each block $\mathcal{B}_i$ consists of the following components:

- Block header, which includes a timestamp, the hash of the previous block, the Merkle root of the transactions, and a nonce.
- Transaction data, represented as an ordered set of transactions.

$$\{tx_1, tx_2, \ldots, tx_m\}.$$

The cryptographic linkage between blocks ensures data integrity, immutability, and resistance to tampering.

Note: A consensus mechanism is a protocol that ensures all nodes in a distributed network agree on the current state of the blockchain.

Smart Contracts

Smart contracts (Buterin, 2014) are self-executing programs deployed on blockchain that automatically enforce predefined rules and conditions. They enable trust less automation of decision-making processes.

9.3 MAIN RESULTS

Section 9.3 presents the main results of the proposed framework, structured into several subsections that capture its core technical contributions. It begins with the neutrosophic blockchain architecture, followed by the neutrosophic score function for evaluating truth, indeterminacy, and falsity. The neutrosophic proof-of-authority (NPoA) is then introduced, along with the neutrosophic multi-criteria decision-making model for enhanced decision accuracy under uncertainty. The section also includes the Smart Contract Implementation of these mechanisms and concludes with a Security Analysis demonstrating improved transparency, robustness, and reliability.

9.3.1 Neutrosophic Blockchain Architecture

We propose a layered architecture for blockchain-based neutrosophic decision systems.

Figure 9.1 illustrates the proposed Neutrosophic Blockchain Architecture, which is organized into five functional layers. The application layer provides user-facing decision interfaces that allow stakeholders to interact with the system. The smart contract layer embeds neutrosophic logic rules, enabling automated execution of decisions that incorporate truth, indeterminacy, and falsity components. The consensus layer is responsible for implementing the neutrosophic consensus mechanism,

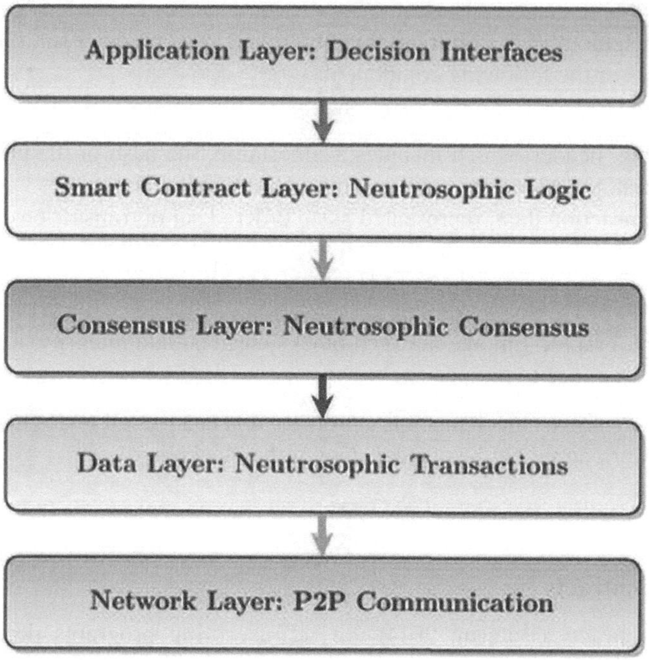

FIGURE 9.1 Neutrosophic Blockchain Architecture.

ensuring that all validator nodes collectively agree on decisions under uncertainty. The data layer manages neutrosophic transactions, securely recording decision evidence and audit trails on the blockchain. At the foundation, the network layer supports peer-to-peer (P2P) communication among nodes, ensuring reliable message exchange and distributed operation. Together, these layers establish a structured and uncertainty aware blockchain framework capable of supporting robust decision-making processes in complex environments.

9.3.2 Neutrosophic Score Function

To facilitate decision-making and ranking, we define a neutrosophic score function:

Definition 4 For an SVNN (Smarandache, 1999; Wang et al., 2010) $\mathcal{N} = \langle \mathcal{T}, \mathcal{I}, \mathcal{F} \rangle$, the score function is defined as:

$$\mathcal{S}(\mathcal{N}) = \frac{2 + \mathcal{T} - \mathcal{I} - \mathcal{F}}{3},$$

where, $\mathcal{S}(\mathcal{N}) \in [0,1]$ and higher score indicates better alternative.

Theorem 1 Let $\mathcal{N}_1 = (\mathcal{T}_1, \mathcal{I}_1, \mathcal{F}_1)$ and $\mathcal{N}_2 = (\mathcal{T}_2, \mathcal{I}_2, \mathcal{F}_2)$ be two SVNNs (Smarandache, 1999; Smarandache and Jdid, 2023; Fujita and Smarandache, 2025). Let $\mathcal{S}$ denotes the score function and $\mathcal{A}$ denotes the accuracy function, defined as

$$\mathcal{A}(\mathcal{N}) = \mathcal{T} - \mathcal{F}.$$

Then, the following ranking rules hold:

1. If $\mathcal{S}(\mathcal{N}_1) > \mathcal{S}(\mathcal{N}_2)$, then $\mathcal{N}_1$ is superior to $\mathcal{N}_2$.
2. If $\mathcal{S}(\mathcal{N}_1) = \mathcal{S}(\mathcal{N}_2)$, then the comparison between $\mathcal{N}_1$ and $\mathcal{N}_2$ is determined using the accuracy function $\mathcal{A}$.

Table 9.1 compares different uncertainty-modeling approaches used in blockchain decision systems. While fuzzy and intuitionistic fuzzy blockchains handle ambiguity and hesitation, they offer limited support for indeterminacy and rely on classical trust mechanisms such as PoS, PoW, or PoA. The proposed neutrosophic blockchain extends these models by incorporating both ambiguity and indeterminacy and introduces an NPoA-based trust mechanism. This broader uncertainty representation makes it suitable for complex domains like healthcare and finance.

TABLE 9.1
Comparison of Uncertainty Modeling Approaches in Blockchain Decision Systems

Approach	Uncertainty Type	Trust Mechanism	Indeterminacy Modeled	Typical Domain
Fuzzy Blockchain	Ambiguity	PoS/PoW	-	IoT, logistics
Intuitionistic Fuzzy Blockchain	Hesitation	PoA	Partial	Supply chain
Proposed Neutrosophic Blockchain	Ambiguity + Indeterminacy	NPoA	Yes	Healthcare, finance

Source: Adapted from Nakamoto (2008) and Buterin (2014).

9.3.3 Neutrosophic Consensus Algorithm

We propose an NPoA consensus mechanism. Table 9.2 shows the neutrosophic consensus algorithm. In this table, all the eight steps of the algorithm are described. The neutrosophic consensus process begins with a network of validator nodes $\mathcal{V} = \{v_1, v_2, \ldots, v_k\}$ that receive a proposed block B together with its associated

TABLE 9.2
Neutrosophic Consensus Algorithm

Step	Description
1	Given a set of validator nodes $\mathcal{V} = \{v_1, v_2, \ldots, v_k\}$ and a proposed block $\mathcal{B}$ with neutrosophic decision $\mathcal{D}$.
2	Each validator v_i evaluates the proposed decision $\mathcal{D}$ and assigns a neutrosophic vote $$\mathcal{N}_i = \langle \mathcal{T}_i, \mathcal{I}_i, \mathcal{F}_i \rangle.$$
3	Validator v_i broadcasts its neutrosophic vote $\mathcal{N}_i$ to all other validators.
4	After receiving all votes, collect the complete set $$\{\mathcal{N}_1, \mathcal{N}_2, \ldots, \mathcal{N}_k\}.$$
5	Compute aggregated neutrosophic values: $$\mathcal{T}_{\text{agg}} = \frac{1}{k}\sum_{i=1}^{k} \mathcal{T}_i,\ \mathcal{I}_{\text{agg}} = \frac{1}{k}\sum_{i=1}^{k} \mathcal{I}_i,\ \mathcal{F}_{\text{agg}} = \frac{1}{k}\sum_{i=1}^{k} \mathcal{F}_i.$$
6	Form the aggregated neutrosophic consensus: $$\mathcal{N}_{\text{consensus}} = \langle \mathcal{T}_{\text{agg}}, \mathcal{I}_{\text{agg}}, \mathcal{F}_{\text{agg}} \rangle.$$
7	Compute the consensus score $\mathcal{S}(\mathcal{N}_{\text{consensus}})$.
8	If $\mathcal{S}(\mathcal{N}_{\text{consensus}}) \geq \theta$, accept the block $\mathcal{B}$; otherwise, reject it.

Source: Adapted from Pramanik, 2022; Smarandache and Jdid, 2023; Parikh and Sahni, 2024; Fujita and Smarandache, 2025.

neutrosophic decision D. Each validator independently analyzes the decision and produces a neutrosophic vote represented as $\mathcal{N}_i = \langle \mathcal{T}_i, \mathcal{I}_i, \mathcal{F}_i \rangle$, which captures its individual degrees of truth, indeterminacy, and falsity. After generating this vote, every validator broadcasts its assessment to all other nodes, ensuring that each participant ultimately obtains the complete set of neutrosophic evaluations.

Once all votes have been collected, the network computes the aggregated neutrosophic values by averaging the truth, indeterminacy, and falsity components across all validators. These aggregated values, denoted as $\mathcal{T}_{\text{agg}}, \mathcal{I}_{\text{agg}}, \mathcal{F}_{\text{agg}}$, form the consensus neutrosophic vector $\mathcal{N}_{\text{consensus}} = \langle \mathcal{T}_{\text{agg}}, \mathcal{I}_{\text{agg}}, \mathcal{F}_{\text{agg}} \rangle$. A consensus score is then calculated from this aggregated vector, and the proposed block is accepted only if the score meets or exceeds a predefined threshold θ; otherwise, the block is rejected. This mechanism ensures a collective and uncertainty-aware decision-making process within the validator network.

9.3.4 Neutrosophic Multi-Criteria Decision-Making

For multi-criteria decision-making (Liu and Wang, 2014; Nabeeh and Tantawy, 2023) with m alternatives and n criteria:

Step 1: Construct neutrosophic decision matrix:

Each alternative is evaluated under each criterion using an SVNN

$$\mathcal{D}=\left[\mathcal{N}_{ij}\right]_{m\times n}=\begin{array}{c} \\ A_1 \\ A_2 \\ \vdots \\ A_m \end{array}\begin{array}{cccc} C_1 & C_2 & \cdots & C_n \\ \langle \mathcal{T}_{11},\mathcal{I}_{11},\mathcal{F}_{11}\rangle & \langle \mathcal{T}_{12},\mathcal{I}_{12},\mathcal{F}_{12}\rangle & \cdots & \langle \mathcal{T}_{1n},\mathcal{I}_{1n},\mathcal{F}_{1n}\rangle \\ \langle \mathcal{T}_{21},\mathcal{I}_{21},\mathcal{F}_{21}\rangle & \langle \mathcal{T}_{22},\mathcal{I}_{22},\mathcal{F}_{22}\rangle & \cdots & \langle \mathcal{T}_{2n},\mathcal{I}_{2n},\mathcal{F}_{2n}\rangle \\ \vdots & \vdots & \ddots & \vdots \\ \langle \mathcal{T}_{m1},\mathcal{I}_{m1},\mathcal{F}_{m1}\rangle & \langle \mathcal{T}_{m2},\mathcal{I}_{m2},\mathcal{F}_{m2}\rangle & \cdots & \langle \mathcal{T}_{mn},\mathcal{I}_{mn},\mathcal{F}_{mn}\rangle \end{array} \quad (9.4)$$

where $A_i\,(i=1,2,\ldots,m)$ denotes the alternatives, $C_j\,(j=1,2,\ldots,n)$ represents the criteria, and $\mathcal{N}_{ij}=\langle \mathcal{T}_{ij},\mathcal{I}_{ij},\mathcal{F}_{ij}\rangle$ is an SVNN corresponding to the evaluation of alternative A_i with respect to criterion C_j.

Step 2: Normalize the decision matrix obtained in Step 1:

In this step, we normalized the decision matrix obtained in Step 1. Normalization depends on whether a criterion is benefit or cost type:

$$\bar{\mathcal{N}}_{ij}^{*}=\begin{cases} \mathcal{N}_{ij}^{*}, & \text{for benefit criteria,} \\ \langle \mathcal{F}_{ij},\mathcal{I}_{ij},\mathcal{T}_{ij}\rangle, & \text{for cost criteria.} \end{cases} \quad (9.5)$$

Here in benefit criteria keep the neutrosophic value as it is (or normalize using the selected method), whereas the cost criteria reverse truth and falsity components.

Step 3: Determine criteria weights:

Let

$$w=\left(w_1,w_2,\ldots,w_n\right),\ \sum_{j=1}^{n} w_j=1. \quad (9.6)$$

The weights may be determined through expert judgment or weight-determination methods (AHP, entropy, BWM, etc.).

Step 4: Compute weighted neutrosophic values:

For each alternative, compute the weighted neutrosophic aggregation across all criteria:

$$\mathcal{N}_i^{w}=\bigotimes_{j=1}^{n}\left(w_j\otimes\bar{\mathcal{N}}_{ij}\right). \quad (9.7)$$

This means that each normalized neutrosophic evaluation is multiplied by its corresponding weight and then aggregated across criteria.

Step 5: Rank alternatives:

Evaluate each alternative using a neutrosophic score function $\mathcal{S}(\mathcal{N}_i^w)$. A common score function is defined as:

$$\mathcal{S}(\mathcal{N}) = \frac{2 + \mathcal{T} - \mathcal{I} - \mathcal{F}}{3}. \tag{9.8}$$

After computing the score for each alternative:

- Sort the alternatives in descending order of score.
- The alternative with the highest score is identified as the best choice.

9.3.5 Smart Contract Implementation

Smart contracts (Smarandache and Jdid, 2023; Fujita and Smarandache, 2025) for neutrosophic decision execution include:

- Neutrosophic voting: collects and aggregates neutrosophic votes.
- Decision executor: automatically executes decisions based on neutrosophic thresholds.
- Audit trail: maintains immutable record of all decisions and their neutrosophic values.

9.3.6 Security Analysis

The proposed neutrosophic consensus mechanism ensures immutability, as any recorded decision cannot be altered without recomputing the entire subsequent chain of blocks. Furthermore, the system is Byzantine fault tolerant, maintaining correct consensus even in the presence of up to $f < \frac{n}{3}$ malicious or faulty validator nodes.

9.3.6.1 Immutability

Once a neutrosophic decision (Smarandache, 1999; Wang et al., 2010; Smarandache and Jdid, 2023; Fujita and Smarandache, 2025) is recorded in a blockchain and confirmed through a consensus process, it becomes computationally infeasible to modify. Any alteration would require recalculating all subsequent blocks, which ensures data integrity and prevents unauthorized manipulation.

9.3.6.2 Byzantine Fault Tolerance

The neutrosophic consensus mechanism (Smarandache, 1999; Wang et al., 2010; Smarandache and Jdid, 2023; Fujita and Smarandache, 2025) is designed to tolerate Byzantine failures. Specifically, the system remains reliable provided that the number of faulty or malicious validator nodes satisfies $f < \frac{n}{3}$, where n denotes the total number of participating nodes.

TABLE 9.3
Comparative Security and Reliability Analysis

Feature	PoA (Mythili and Venkataraman, 2020)	PBFT (Castro and Liskov, 1999)	Proposed NPoA
Fault tolerance	$f < \frac{n}{3}$	$f < \frac{n}{3}$	$f < \frac{n}{3}$ with uncertainty adaptation
Transparency	Medium	High	High + Neutrosophic audit trail
Uncertainty handling	None	Limited (probabilistic)	Explicit $(\mathcal{T}, \mathcal{I}, \mathcal{F})$
Decision latency	Low	Moderate	Moderate
Trust source	Authority nodes	Byzantine agreement	Neutrosophic evaluation nodes

Table 9.3 presents a comparative assessment of consensus mechanisms from the perspective of security, reliability, and uncertainty-aware decision support. Both PoA and PBFT maintain classical Byzantine fault tolerance thresholds of $f < \frac{n}{3}$; however, PoA provides only moderate transparency and lacks any uncertainty modeling capability. PBFT enhances transparency and incorporates limited probabilistic reasoning but does not explicitly represent indeterminate states. The proposed NPoA framework preserves the same fault tolerance bound while introducing uncertainty-adaptive behavior, enabling the system to respond dynamically to fluctuating or incomplete information.

In addition, NPoA integrates a neutrosophic audit trail, offering high interpretability by capturing truth $(\mathcal{T})$, indeterminacy $(\mathcal{I})$, and falsity $(\mathcal{F})$ values during the validation process. Decision latency remains moderate and comparable to Practical Byzantine Fault Tolerance (PBFT), but trust is no longer dependent on fixed authority nodes or traditional Byzantine agreement. Instead, it is delegated to neutrosophic evaluation nodes capable of performing granular uncertainty assessments, thereby improving resilience and decision accuracy under complex or ambiguous operational environments.

9.4 APPLICATIONS

Section 9.4 presents the key applications that demonstrate the practical utility of the proposed framework. The first subsection focuses on supply chain management, illustrating how neutrosophic modeling enhances traceability, transparency, and decision reliability under uncertain conditions. The subsequent subsections explore healthcare decision support and financial risk assessment, highlighting the framework's capability to manage ambiguity and incomplete information in critical domains. Finally, the performance evaluation subsection assesses the effectiveness of the proposed methods across multiple operational scenarios, confirming their robustness and adaptability in real-world environments.

9.4.1 Supply Chain Management

In supply-chain decision-making (Ahmad, 2022; Aytekin et al., 2023; Nafei et al., 2024), neutrosophic values provide a structured way to capture multiple dimensions of uncertainty. The truth component $(\mathcal{T})$ reflects the assessed reliability of a supplier based on past performance and consistency. The indeterminacy component $(\mathcal{I})$ represents the uncertainty associated with delivery times, which may arise from fluctuating logistics conditions or incomplete information. The falsity component $(\mathcal{F})$ indicates the potential risk of quality issues that could affect product integrity or customer satisfaction. Together, these neutrosophic parameters enable a more comprehensive and realistic evaluation of supplier performance in uncertain environments. Blockchain ensures transparency and traceability of all supply chain decisions.

9.4.1.1 Illustrative Example

A numerical example is presented to demonstrate the applicability of the single-valued neutrosophic framework in supplier evaluation within a supply chain management context. Supplier selection is inherently characterized by incomplete information, conflicting expert opinions, and uncertain operational conditions. To address these challenges, neutrosophic logic is employed to simultaneously capture reliability, uncertainty, and risk.

Consider two candidate suppliers, denoted by $\mathcal{S}_1$ and $\mathcal{S}_2$, which are evaluated based on three decision attributes: supplier reliability (truth-membership, $\mathcal{T}$), delivery-related uncertainty (indeterminacy-membership, $\mathcal{I}$), and the risk of quality nonconformance (falsity-membership, $\mathcal{F}$). The assessment values are obtained from historical performance data, expert evaluations, and verified supply chain records.

The neutrosophic representations of the suppliers are expressed as

$$\mathcal{N}_{S_1} = \left\langle \mathcal{T}_{S_1}, \mathcal{I}_{S_1}, \mathcal{F}_{S_1} \right\rangle = \langle 0.8,\ 0.1,\ 0.2 \rangle$$

$$\mathcal{N}_{S_2} = \left\langle \mathcal{T}_{S_2}, \mathcal{I}_{S_2}, \mathcal{F}_{S_2} \right\rangle = \langle 0.6,\ 0.2,\ 0.3 \rangle$$

A higher value of $\mathcal{T}$ indicates stronger supplier reliability, whereas higher values of $\mathcal{I}$ and $\mathcal{F}$ correspond to increased delivery uncertainty and quality-related risk, respectively.

To rank the suppliers, the following neutrosophic score function is adopted:

$$S(\mathcal{N}) = \frac{2 + \mathcal{T} - \mathcal{I} - \mathcal{F}.}{3}$$

Using this formulation, the score values for the two suppliers are computed as

$$S\left(\mathcal{N}_{S_1}\right) = 0.83, \quad S\left(\mathcal{N}_{S_2}\right) = 0.70.$$

Since $S\left(\mathcal{N}_{S_1}\right) > S\left(\mathcal{N}_{S_2}\right)$, supplier $\mathcal{S}_1$ is ranked higher and is therefore selected as the preferred supplier. The result indicates that $\mathcal{S}_1$ exhibits superior overall performance, characterized by higher reliability and lower levels of uncertainty and quality risk.

This illustrative example demonstrates that neutrosophic modeling provides a more comprehensive and realistic decision-support mechanism than classical fuzzy or deterministic approaches, as it explicitly incorporates indeterminacy alongside positive and negative assessments. Consequently, the proposed framework is well suited for supplier evaluation problems in modern supply chain systems operating under uncertain and dynamic conditions.

9.4.2 Healthcare Decision Support

In medical diagnosis, neutrosophic values (Ahmed et al., 2021; Ihsan et al., 2023; Guerrero et al., 2023; Sajid et al., 2025) offer a systematic way to capture the complexities of clinical uncertainty. The truth component $(\mathcal{T})$ represents the level of confidence a clinician has in a particular diagnosis based on observable symptoms and established medical indicators. The indeterminacy component $(\mathcal{I})$ reflects the presence of unclear or partially expressed symptoms that do not strongly support or refute any specific condition. The falsity component $(\mathcal{F})$ accounts for contradictory or conflicting test results that challenge the initial diagnostic assumption. By integrating these three dimensions, neutrosophic modeling helps provide a more balanced and realistic assessment of diagnostic outcomes under uncertainty. Blockchain maintains secure, auditable medical decision records. This approach ensures that diagnostic reasoning is traceable and that all degrees of uncertainty are systematically documented, ultimately helping improve the quality and reliability of medical decisions.

9.4.2.1 Illustrative Example

To illustrate the applicability of neutrosophic modeling in healthcare decision-support systems, consider a medical diagnosis problem characterized by uncertain, incomplete, and occasionally conflicting clinical information. Accurate diagnosis in such environments requires a framework capable of simultaneously representing confidence, ambiguity, and contradiction in clinical evidence.

Assume that a clinician evaluates two possible diagnostic hypotheses for a patient presenting with respiratory symptoms, denoted by $\mathcal{D}_1$ (e.g., bacterial pneumonia) and $\mathcal{D}_2$ (e.g., viral respiratory infection). Each diagnosis is assessed using a single-valued neutrosophic set based on three components: diagnostic confidence (truth-membership, $\mathcal{T}$), clinical indeterminacy (indeterminacy-membership, $\mathcal{I}$), and contradictory medical evidence (falsity-membership, $\mathcal{F}$). The assessments are derived from clinical examination, laboratory investigations, imaging reports, and expert judgment, with all records securely stored in a blockchain-based medical information system.

The neutrosophic representations of the diagnostic hypotheses are given by

$$\mathcal{N}_{D_1} = \left\langle \mathcal{T}_{D_1}, \mathcal{I}_{D_1}, \mathcal{F}_{D_1} \right\rangle = \langle 0.75, 0.15, 0.20 \rangle$$

$$\mathcal{N}_{D_2} = \left\langle \mathcal{T}_{D_2}, \mathcal{I}_{D_2}, \mathcal{F}_{D_2} \right\rangle = \langle 0.55, 0.25, 0.30 \rangle$$

Here, a higher value of $\mathcal{T}$ indicates stronger clinical support for a diagnosis based on observed symptoms and test results. The indeterminacy component $\mathcal{I}$ represents ambiguous or partially expressed clinical signs, while the falsity component

$\mathcal{F}$ reflects conflicting laboratory findings or imaging results that contradict the diagnosis.

To evaluate and rank the diagnostic hypotheses, the following neutrosophic score function is employed:

$$S(\mathcal{N}) = \frac{2 + \mathcal{T} - \mathcal{I} - \mathcal{F}}{3}$$

Using this function, the diagnostic scores are computed as

$$S(\mathcal{N}_{D_1}) = 0.80, \quad S(\mathcal{N}_{D_2}) = 0.67$$

Since $S(\mathcal{N}_{D_1}) > S(\mathcal{N}_{D_2})$, diagnostic hypothesis $\mathcal{D}_1$ is considered more plausible under the given clinical evidence. This result suggests that bacterial pneumonia is the more likely diagnosis, although a degree of uncertainty and contradiction remains.

This example demonstrates that neutrosophic modeling enables clinicians to explicitly incorporate indeterminacy and conflicting evidence into diagnostic reasoning, which is not adequately addressed by conventional fuzzy or probabilistic approaches. Furthermore, the integration of blockchain technology ensures that all diagnostic evaluations and decision pathways are securely recorded, traceable, and auditable, thereby enhancing transparency, accountability, and reliability in healthcare decision-support systems.

9.4.3 Financial Risk Assessment

In investment decision analysis, neutrosophic risk assessment (Metawa and Mourad, 2022; Zhang et al., 2023; Nguyen et al., 2024; Lin and Lo, 2024) provides a systematic framework for addressing financial uncertainty. The truth membership component $(\mathcal{T})$ represents the likelihood of achieving expected returns based on historical performance, financial indicators, and market trends. The indeterminacy component $(\mathcal{I})$ captures uncertain and ambiguous factors such as economic volatility, incomplete information, and unforeseen macroeconomic changes that may influence asset behavior. The falsity membership component $(\mathcal{F})$ reflects the potential for losses resulting from adverse market movements and unexpected financial risks. By integrating these three components, the neutrosophic approach enables a more comprehensive and flexible evaluation of investment alternatives under dynamic market conditions. In addition, smart contracts support the automated implementation of risk-aware investment strategies derived from neutrosophic evaluations, thereby enhancing transparency, consistency, and efficiency in decision execution.

9.4.3.1 Illustrative Example

In an investment decision-making environment characterized by uncertainty and market volatility, two investment options, denoted by A_1 and A_2, are considered. To account for imprecise and incomplete financial information, each option is represented using a single-valued neutrosophic set. The truth-membership degree $(\mathcal{T})$ reflects the confidence in achieving the expected return, the indeterminacy-membership degree $(\mathcal{I})$ represents uncertain economic conditions and incomplete market signals, and the falsity-membership degree $(\mathcal{F})$ denotes the possibility of financial loss. Based on historical trends and expert assessment, the neutrosophic evaluations are expressed as

$$\mathcal{N}_{A_1} = \left\langle \mathcal{T}_{A_1}, \mathcal{I}_{A_1}, \mathcal{F}_{A_1} \right\rangle = \langle 0.70, 0.20, 0.25 \rangle$$

$$\mathcal{N}_{A_2} = \left\langle \mathcal{T}_{A_2}, \mathcal{I}_{A_2}, \mathcal{F}_{A_2} \right\rangle = \langle 0.55, 0.15, 0.30 \rangle$$

Applying the score function

$$\mathcal{S}(\mathcal{N}) = \frac{2 + T - I - F}{3},$$

Using this function, the diagnostic scores are computed as

$$S\left(\mathcal{N}_{A_1}\right) = 0.75, \quad S\left(\mathcal{N}_{A_2}\right) = 0.70$$

Hence, as $S(A_1) > S(A_2)$, the investment option A_1 is preferred and demonstrates a comparatively lower-risk profile within the neutrosophic decision framework. This brief illustration highlights the applicability of neutrosophic modeling for financial risk assessment under uncertain decision environments.

The obtained results indicate that the incorporation of neutrosophic uncertainty modeling into blockchain consensus mechanisms enhances both the quantitative performance metrics and the qualitative trust attributes of distributed decision-making systems.

9.5 CONCLUSION AND FUTURE DIRECTIONS

This paper presents a comprehensive framework for blockchain-based decision systems operating under neutrosophic uncertainty. By integrating neutrosophic logic with blockchain technology, the proposed approach effectively addresses key challenges in distributed decision-making, including uncertainty quantification, transparency, and trust management. The introduced neutrosophic consensus mechanism enables reliable decision-making in the presence of incomplete, inconsistent, or indeterminate information. Furthermore, the use of smart contracts facilitates the automated execution of neutrosophic decisions while ensuring full traceability and auditability. The results indicate that the proposed framework enhances decision quality in uncertain environments without compromising the inherent security and transparency properties of blockchain technology. Owing to its flexible uncertainty modeling capability, the framework is applicable to a wide range of domains, including supply chain management, healthcare decision-support, and financial systems. Future research may focus on extending the proposed framework to large-scale and heterogeneous blockchain networks, investigating performance optimization under high transaction volumes, and integrating machine learning techniques for dynamic neutrosophic parameter estimation. Additionally, real-world pilot implementations and comparative studies with other advanced consensus mechanisms will further validate the scalability, robustness, and practical applicability of the proposed approach.

REFERENCES

Ahmad, F.: Interactive neutrosophic optimization technique for multi objective programming problems: An application to pharmaceutical supply chain management. Annals of Operations Research 311(2), 551–585 (2022).

Ahmed, R., Nasiri, F., Zayed, T.: A novel neutrosophic-based machine learning approach for maintenance prioritization in healthcare facilities. Journal of Building Engineering 42, 102480 (2021).

Aliahmadi, A., Nozari, H.: Evaluation of security metrics in AIoT and block chain based supply chain by neutrosophic decision-making method. Supply Chain Forum: An International Journal 24, 31–42 (2023).

Aytekin, A., Okoth, B.O., Korucuk, S., Karamaşa, Ç., Tirkolaee, E.B.: A neutrosophic approach to evaluate the factors affecting performance and theory of sustainable supply chain management: Application to textile industry. Management Decision 61(2), 506–529 (2023).

Bardhan, R., Chauhan, D., Sahni, M.: A neutrosophic topsis framework for multicriteria selection of functionally graded materials. Procedia Structural Integrity 72, 507–519 (2025).

Buterin, V.: A next-generation smart contract and decentralized application platform. White Paper 3(37), 2–1 (2014).

Cali, U., Deveci, M., Saha, S.S., Halden, U., Smarandache, F.: Prioritizing energy blockchain use cases using type-2 neutrosophic number-based EDAS. IEEE Access 10, 34260–34276 (2022).

Castro, M., Liskov, B.: Practical byzantine fault tolerance. In: Proceedings of the Third Symposium on Operating Systems Design and Implementation, New Orleans, USA, 99, pp. 173–186 (1999).

Fujita, T., Smarandache, F.: Local-neutrosophic logic and local-neutrosophic sets: Incorporating locality with applications. Multicriteria Algorithms with Applications 6, 68–86 (2025).

Guerrero, O.E.A., Bandera, N.D.L.M.H., Bernal, L.I.M., Espinosa, J.S.P.: Neutrosophic modeling of healthcare resource allocation. Neutrosophic Sets and Systems 62, 60–68 (2023).

Ihsan, M., Saeed, M., Alanzi, A.M., Khalifa, H.E.W.: An algorithmic multiple attribute decision-making method for heart problem analysis under neutrosophic hyper soft expert set with fuzzy parameterized degree-based setting. Peer Journal of Computer Science 9, 1607 (2023).

Lin, S.W., Lo, H.W.: An FMEA model for risk assessment of university sustainability: Using a combined ITARA with TOPSIS-AL approach based neutrosophic sets. Annals of Operations Research 342(3), 2119–2145 (2024).

Liu, P., Wang, Y.: Multiple attribute decision-making method based on single valued neutrosophic normalized weighted Bonferroni mean. Neural Computing and Applications 25(7), 2001–2010 (2014).

Metawa, N., Mourad, N.: Neutrosophic-based multi-objectives model for financial risk management. International Journal of Neutrosophic Science (IJNS) 19(1) (2022).

Mythili, R., Venkataraman, R.: Proof of policy (pop): A new attribute-based blockchain consensus protocol. In: V. Singh, V. K. Asari, S. Kumar, & R. B. Patel (Eds.), Computational Methods and Data Engineering: Proceedings of ICMDE 2020, Volume 1, pp. 451–464. Springer, Singapore (2020). https://doi.org/10.1007/978-981-15-6876-3_35

Nabeeh, N.A., Tantawy, A.A.: A neutrosophic model for blockchain platform selection based on SWARA and WSM. Neutrosophic and Information Fusion 1(2), 29–43 (2023).

Nafei, A., Azizi, S.P., Edalatpanah, S.A., Huang, C.Y.: Smart topsis: A neural network-driven topsis with neutrosophic triplets for green supplier selection in sustainable manufacturing. Expert Systems with Applications 255, 124744 (2024).

Nakamoto, S.: Bitcoin: A peer-to-peer electronic cash system. Available at SSRN 3440802 (2008). https://assets.pubpub.org/d8wct41f/31611263538139.pdf

Nguyen, P.H., Nguyen, L.A.T., Pham, H.A.T., Nguyen, T.H.T., Vu, T.G.: Assessing cybersecurity risks and prioritizing top strategies in Vietnam's finance and banking system using strategic decision-making models-based neutrosophic sets and z number. Heliyon 10(19) (2024).

Parikh, M., Sahni, M.: Sumudu Transform for Solving Second Order Ordinary Differential Equation Under Neutrosophic Initial Conditions vol. 38. Infinite Study, New Mexico (2020).

Parikh, M., Sahni, M.: Solution of logistic differential equation in an uncertain environment using neutrosophic numbers. Journal of Interdisciplinary Mathematics 27, 145–169 (2024).

Parikh, M., Sahni, M., Sahni, R.: Solution of first order initial value problem using analytical and numerical method in neutrosophic environment. Neutrosophic Sets and Systems 51(1), 20 (2022).

Pramanik, S.: Single-valued neutrosophic set: An overview. Transdisciplinarity, 563–608 (2022).

Sajid, M., Khan, K.A., Frnda, J., Rahman, A.U.: A novel multi-attribute decision making method for supplier selection in the health care industry using cosine similarity measures of single-valued neutrosophic cubic hyper soft sets. IEEE Access (2025).

Smarandache, F.: "Neutrosophy: Neutrosophic probability, set, and logic"-first version. Chapter Volume III (1999).

Smarandache, F., & Jdid, M. (2023). On overview of neutrosophic and plithogenic theories and applications. Prospects for Applied Mathematics and Data Analysis, 2(1), 19–26. https://doi.org/10.54216/PAMDA.020102

Vafadarnikjoo, A., Badri Ahmadi, H., Liou, J.J., Botelho, T., Chalvatzis, K.: Analyzing blockchain adoption barriers in manufacturing supply chains by the neutrosophic analytic hierarchy process. Annals of Operations Research 327(1), 129–156 (2023).

Wang, H., Smarandache, F., Zhang, Y., Sunderraman, R.: Single Valued Neutrosophic Sets, pp. 410–413. Infinite Study, United States (2010).

Ye, J.: A multicriteria decision-making method using aggregation operators for simplified neutrosophic sets. Journal of Intelligent & Fuzzy Systems 26(5), 24592466 (2014).

Zhang, H., Zhang, Y., Rahman, A.U., Saeed, M.: An intelligent SV-neutrosophic parameterized MCDM approach to risk evaluation based on complex fuzzy hypersoft set for real estate investments. Management Decision 61(2), 486–505 (2023).

10 Penalty-Based Bonferroni Means

A Heterogeneity Penalty Function Based on Bonferroni OWA Operator and Adequacy Coefficient

Fabio Blanco-Mesa

10.1 INTRODUCTION

Aggregation operators and aggregation functions have developed many methods and extensions that have been applied in decision-making processes in different fields (Blanco-Mesa et al., 2019). The present research has focused on the development of aggregation tools that integrate various types of information to obtain a single representative value (Beliakov et al., 2007, p. 200, 2016). In this context, research related to the Bonferroni mean (Bonferroni, 1950) is noteworthy, as it allows for multiple comparisons, which guarantees the compensation of possible errors through the alpha significance level between each comparison (Yager, 2009) compensating for possible errors through the alpha significance level between each comparison.

In light of these characteristics, a plethora of approaches have been developed, including generalized (Yager, 2009), geometric (Xia et al., 2012), Pythagorean (Liang et al., 2019), triangular fuzzy numbers (Li & Wang, 2019), OWA operator (Alfaro-Garcia et al., 2022; Janani et al., 2023), hesitant fuzzy set (Ren et al., 2022), intuitionistic (Liu et al., 2023), linguistic (Yang et al., 2024), interval-valued (Ali & Rasool, 2024), probabilistic (Espinoza-Audelo et al., 2020), dual form (Liu & Liu, 2021), and so on. Among these, the OWA operator is of particular interest as a method of reordering information that considers specific ranges of minimum and maximum values to obtain results (Baez-Palencia et al., 2019; Yager, 1988). A number of extensions have been proposed in combination with the Bonferroni mean to address issues in administrative sciences with induced variable characteristics, heavyweight vectors, distance measures, probability (Espinoza-Audelo et al., 2020) and prioritization (Perez-Arellano et al., 2020).

Notwithstanding the operator's versatility, the implications of the decision maker's disagreement and consensus have not yet been given due consideration. In this sense Beliakov et al. (2011) and Calvo and Beliakov (2010) proposed penalty-based

DOI: 10.1201/9781003724902-11

aggregation functions to address disagreement and consensus in decision-making problems. As posited by Beliakov et al. (2011), the utilization of aggregation functions facilitates the resolution of decision-making challenges by encompassing diverse distributions, data asymmetry, and expert insights. This approach enhances the analytical or numerical optimization of processes, while simultaneously facilitating a more nuanced interpretation of consensus by accounting for individual discrepancies, interaction, heterogeneity, and significance.

In consideration of the previously mentioned points, the present research puts forward a number of novel heterogenous penalty aggregation functions. These functions are referred to as the penalty weight adequacy coefficient (P–WAC), the double penalty weight adequacy coefficient (dP–WAC), and the Bonferroni penalty adequacy coefficient OWA (BP_{AC}–OWA). These novel operators present a new heterogenous penalty using the adequacy coefficient (AC) and a new extension of the Bonferroni-OWA mean family, which facilitates the capture of the importance, expectations, and disagreements inherent in decision-making processes. Moreover, the text encompasses the methodology for the construction of the new operator, its characteristics, properties, generalized and quasi-arithmetic means. The chapter concludes with the presentation of a mathematical example.

The chapter is structured as follows: Section 10.2 presents basic definitions of the penalty function, AC, OWA, and BON-OWA operator. Section 10.3 presents the new heterogenous penalty-based function. Section 10.4 presents penalty-based BON-OWA in combination with aggregation operators. Section 10.5 presents penalty-based PB_{AC}-OWA in combination with generalized and quasi-arithmetic aggregation operators. Finally, Section 10.6 presents mathematical example and Section 10.7 presents conclusions of the work.

10.2 PRELIMINARIES

In this section, we present the main concepts of penalty function, AC, OWA, and BON-OWA operator to develop new based heterogeny penalty function aggregation.

10.2.1 Penalty Function

The penalty function is a measure of disagreement or dissimilarity that has been adjusted to allow for the minimization of penalties in the aggregation of expert opinions (Calvo & Beliakov, 2010). This function is defined as follows:

Definition 1. The function $P: X^{n+1} \to \overline{\mathbb{R}}_+ = [0, \infty)$ is a penalty function if and only if it satisfies:

i. $P(x_i, y) \geq 0$ *for all* $i = 1 \ldots n$;
ii. $P(x_i, y) = 0$ *if* $x_i = y_i$ and
iii. For any fixed $\mathbf{x}$, the set of minimizers of $P(\mathbf{x}, y)$ is either an individual or an interval.

Then the penalty-based function is

$$f(x) = \text{argmin}_y P(\mathbf{x}, y), \tag{10.1}$$

where $\mathbf{x} = (x_1, x_2 \ldots x_n)$ is an input vector while y is an output value or a fused value. Additionally, there are two conditions: (1) Negative penalties are not allowed; (2) If there is full agreement, no penalty is imposed.

Thus, penalty-based function, Calvo and Beliakov (2010) proposed penalty-based aggregation function, where inputs are symmetric with penalties saving weights. Then, formulation of penalty-weighted aggregation follows:

$$P(\mathbf{x}, y) = \sum_{i=1}^{n} p_i (x_i - y), \tag{10.2}$$

where p_i absorbed weights.

Thus,

$$P(\mathbf{x}, y) = \sum_{i=1}^{n} w_i |x_i - y|^i, \tag{10.3}$$

where $|x_i - y|^i$ influences the variation of the contribution of the ith input and the weights influence the relative importance of the ith input.

Also, Beliakov et al. (2011) suggest that exits different distributions in the expert opinions. In their example, consider that arguments and penalties have a Laplace and normal distribution, then penalty function is defined as:

$$P(\mathbf{x}, y) = w_1 |x_1 - y| + w_2 (x_2 - y)^2. \tag{10.4}$$

This approximation offers a wide variety of solutions when opinions are very heterogenous and asymmetric.

10.2.2 Distance Measures

The AC (Kaufmann & Gil-Aluja, 1986, 1987) is a technique that allows calculating the comparison between two arguments x_i and y_i and two sets A and B, where it neutralizes the result when the comparison shows that the real element is higher than the ideal one. For two sets A and B, the weighted AC can be defined as follows.

Definition 2. A weighted AC of dimension n is a mapping K: $[0,1]^n \times [0,1]^n \rightarrow [0,1]$ that has an associated weighting vector W of dimension n with $w_j \in [0, 1]$ and $\sum_{j=1}^{n} w_j = 1$:

$$AC((x_1,y_1),\ldots,(x_n,y_n)) = \sum_{i=1}^{n} w_i\left[1\wedge(1-x_i+y_i)\right], \tag{10.5}$$

where x_i and y_i are the *i*th arguments of the sets *X* and *Y*. The neutralization of the result can occur when the comparison shows that the real element is superior to the ideal, whether it tends to 1 or 0, i.e., if $y_i \geq x_i$, then $K(y_i \to x_i) = 1$, which presented in (10.4) or $K(y_i \to x_i) = 0$, which is presented as $\left[0 \vee (x_i - y_i)\right]$.

10.3 NEWS HETEROGENOUS PENALTY-BASED FUNCTION

This study draws inspiration from Belyakov's (Beliakov et al., 2011) contributions to the field of heterogenous penalties, particularly his work on asymmetry and variations in the distributions of expert opinions. Building upon these foundations, two novel heterogenous penalty functions are proposed, utilizing the AC and the maximum and minimum level index. Both proposals are presented below.

A new heterogenous penalty-based aggregation function called penalty weight adequacy coefficient (P–WAC) is proposed. The inputs are symmetric with penalties saving weights, and the penalty is neutralized when *y* is superior to the x_i. The formulation of P–WAC follows:

Proposition 1. A penalty weight adequacy coefficient has an associated weighting vector W of dimension *n* with the sum of the weights 1 and $w_j \in [0,1]$, such that:

$$P-WAC(\mathbf{x},y) = \sum_{i=1}^{n} w_i p\left[1\wedge(1-x_i+y)\right], \tag{10.6}$$

where x_i is the *i*th arguments of the set *X* and y is fused value. Here $\left[1\wedge(1-x_i+y)\right]$ influences the variation of the contribution of the *i*th input and the weights influence the relative importance of the *i*th input. If $y \geq x_i$, then $K(y \to x_i) = 1$. Thus, if $y > x_i$ or $K(y \to x_i) = \left[1\wedge(1+x_i-y)\right]$.

In addition, heterogenous penalties can be used when comparing characteristics of the elements to their ideal. This approach is intended to address more complex aggregations, characterized by greater variations in distribution and increased asymmetry. In this sense, x_i can take the values of distance measurements AC. These comparison values are expressed as $k_D = \mathbf{x}$, where $k_{AC} = \left[1\wedge(1-x_i+y)\right]$. The primary characteristic of this evaluation is that there is a greater penalty for being far from the ideal and for lacking a condition for fulfillment, resulting in a double penalty. Then, $dP(k_D, y) \geq 0$ *for all* $i = 1\ldots n; dP(x_i, y) = 0$ *if* $k_D = y$ and for any fixed k_D, the set of minimizers of $dP(k_D,y)$ is either an individual or an interval. Therefore, $f(k_D)=\arg P_2(k_D,y)$. Thus, this formulation can be expressed as $dP(k_D,y)=\Sigma_{i=1}^{n} p_i\left(\left[1\wedge(1-x_i+y)\right]-y\right)$, where p_i represents absorbed weights. The new formulations for $dP-WAC$ is presented below:

$$dP-WAC(k_D,y)=\sum_{i=1}^{n} w_i p\left[1\wedge(1-k_D+y).\right] \tag{10.7}$$

10.4 PENALTY-BASED BON-OWA IN COMBINATION WITH AGGREGATION OPERATORS

The penalty proposition presented in Section 10.3 can be extended by combining it with penalty aggregation functions (Calvo & Beliakov, 2010) and the OWA and BON-OWA operators (F. Blanco-Mesa, León-Castro, et al., 2019; Leon-Castro et al., 2020). The main advantage is that it can combine penalty-based functions, classical OWA and Bonferroni aggregation in the same formulation. The basic formulations of the OWA operator and the Bonferroni OWA are presented below.

10.4.1 OWA Operators

The OWA operator (Yager, 1988) is a method that allows for the aggregation of information, thereby enabling the calculation of both the maximum and minimum operators. Additionally, it provides a parameterized class of mean-type aggregation operators. The following definition provides a clear understanding of its functionality:

Definition 3. An OWA operator of dimension n is a mapping OWA: $R^n \to R$ that has an associated weighing vector W of dimension n with $w_j \in [0, 1]$ and $\Sigma_{j=1}^{n} w_j = 1$, such that:

$$OWA(a_1,a_2,\ldots,a_n)=\sum_{j=1}^{n} w_j b_j, \tag{10.8}$$

where b_i is the jth largest of the a_i.

10.4.2 Bonferroni Means, BON-OWA

The Bonferroni mean (Bonferroni, 1950) is an averaging aggregation function that allows capturing the interrelationship between arguments such that the formula is:

Definition 4. The Bonferroni means is a mean type of continuous aggregation operator that can be defined as follows:

$$B(a_1,a_2,\ldots,a_n)=\left(\sum_{k=1}^{n}a_i^r\left(\frac{1}{1-n}\sum_{\substack{j=1\\j\neq i}}^{n}a_j^q\right)\right)^{\frac{1}{r+q}}. \tag{10.9}$$

BON-OWA (Yager, 2009) allows aggregating information and making multiple-comparison between input arguments and their interrelationship.

Definition 5. The Bonferroni OWA (BON-OWA) is a mean type of continuous aggregation operator that can be defined as follows:

$$BON-OWA(a_1,\ldots,a_n)=\left(\frac{1}{n}\sum_i a_i^r OWA_W\left(V^i\right)\right)^{\frac{1}{r+q}}, \tag{10.10}$$

where $OWA_W\left(V^i\right)=\left(\frac{1}{\mathrm{n}-1}\Sigma_{\substack{j=1\\j\neq i}}^{\mathrm{n}}\,\mathrm{a}_j^q\right)$ with $\left(V^i\right)$ being the vector of all a_j except a_i and w being an $n-1$ vector W_i associated with α_i whose components w_{ij} are the OWA weights. Let W be an OWA weighing vector of dimension $n-1$ with components $w_i \in [0,1]$ when $\Sigma_i\, w_i = 1$. Then, we can define this aggregation as $OWA_W\left(V^i\right)=\left(\Sigma_{\mathrm{j}=1}^{n-1}\, w_i a_{\pi_k(j)}\right)$, where $a_{\pi_k(j)}$ is the largest element in the tuple V^i and $w_i=\left(\frac{1}{n-1}\right)$ for all i.

Then, using the basic formulations and the proposed penalty functions, improve the way to address the complexity by considering disagreement or dissimilarity, and the attitudinal characteristics of expert opinions, within the decision-making process. Please find the formulations presented below.

Proposition 2. A Bonferroni penalty adequacy coefficient OWA (BP_{AC}–OWA) is a mean type of continuous aggregation operator, that is given by:

$$BP_{AC}-OWA(\mathbf{x},\mathrm{y})=\left(\frac{1}{n}\sum_i P_{ACi}^{\ r}P_{AC}OWA_W\left(V^i\right)\right)^{\frac{1}{r+q}}, \tag{10.11}$$

where $P_{AC}OWA_W\left(V^i\right)=\left(\frac{1}{n-1}\sum_{\substack{j=1\\j\neq i}}^{n} p\left[1\wedge\left(1-x_j+y\right)\right]_j^q\right)$ with $\left(V^i\right)$ being the vector of all $p\left[1\wedge\left(1-x_j+y\right)\right]$ except $p\left[1\wedge\left(1-x_i+y\right)\right]$ and w being an $n-1$ vector W_i associated with α_i whose components w_{ij} are the OWA weights. Let W be an OWA weighing vector of dimension $n-1$ with components $w_i\in[0,1]$ when $\sum_i w_i=1$. Then, we can define this aggregation as $P_{AC}OWA_W\left(V^i\right)=\left(\sum_{j=1}^{n-1} w_i p\left[1\wedge\left(1-x_j+y\right)\right]_{\pi_k(j)}\right)$, where $p\left[1\wedge\left(1-x_j+y\right)\right]_{\pi_k(j)}$ is the largest element in the tuple V^i and $w_i=\left(\frac{1}{n-1}\right)$ for all i.

Proposition 3. A Bonferroni double–penalty adequacy coefficient OWA (dBP$_{AC}$–OWA) is a mean type of continuous aggregation operator, that is given by:

$$dBP_{AC}-OWA\left(k_D,y\right)=\left(\frac{1}{n}\sum_i dP_{AC_i}^{\,r} dP_{AC}OWA_W\left(V^i\right)\right)^{\frac{1}{r+q}}, \tag{10.12}$$

where $dP_{AC}OWA_W\left(V^i\right)=\left(\frac{1}{n-1}\sum_{\substack{j=1\\j\neq i}}^{n} dp\left[1\wedge\left(1-k_D+y\right)\right]_j^q\right)$ with $\left(V^i\right)$ being the vector of all $dp\left[1\wedge\left(1-k_{Dj}+y\right)\right]$ except $dp\left[1\wedge\left(1-k_{Di}+y\right)\right]$ and w being an $n-1$ vector W_i associated with α_i whose components w_{ij} are the OWA weights. Let W be an OWA weighing vector of dimension $n-1$ with components $w_i\in[0,1]$ when $\sum_i w_i=1$. Then, we can define this aggregation as $dP_{AC}OWA_W\left(V^i\right)=\left(\sum_{j=1}^{n-1} w_i dp\left[1\wedge\left(1-k_{Dj}+y\right)\right]_{\pi_k(j)}\right)$, where $dp\left[1\wedge\left(1-k_{Dj}+y\right)\right]_{\pi_k(j)}$ is the largest element in the tuple V^i and $w_i=\left(\frac{1}{n-1}\right)$ for all i.

Let us discuss the properties of BP$_{AC}$–IOWA, BP$_{IMAM}$–IOWA, dBP$_{AC}$–IOWA and dBP$_{IMAM}$–IOWA. Note that the proofs are trivial and thus omitted.

Theorem 1. (Commutativity-OWA and AC. Assume f is BP$_{AC}$–OWA, then

$$f\left(\left(ac_1,y\right),\ldots,\left(ac_n,y\right)\right)=f\left(\left(ac_1,y\right),\ldots,\left(ac_n,y\right)\right). \tag{10.13}$$

$$f\left(\left(k_{D1},y\right),\ldots,\left(k_{Dn},y\right)\right)=f\left(\left(k_{D1},y\right),\ldots,\left(k_{Dn},y\right)\right). \tag{10.14}$$

Theorem 2. (Monotonicity). Assume f is the BP_{AC}–OWA operators; if $[1 \wedge (1 - x_i + y)] \geq [1 \wedge (1 - c_i + y)]$ for all i_i, then

$$f((x_1, y), \ldots, (x_n, y)) \geq f((c_1, y), \ldots, (c_n, y)), \tag{10.15}$$

for both double penalization dBP_{AC}–OWA if $[1 \wedge (1 - k_{Di} + y)] \geq [1 \wedge (1 - l_{Di} + y)]$ for all i_i, then $f((k_{D1}, y), \ldots, (k_{Dn}, y)) \geq f((l_{D1}, y), \ldots, (l_{Dn}, y))$.

Theorem 3. (Bounded). Assume f is the BP_{AC}–IOWA operator, then:

$$\begin{aligned} min\{[1 \wedge (1 - x_i + y)]\} &\leq f((u_1, x_1, y), \ldots, (u_n, x_n, y)) \\ &\leq max\{[1 \wedge (1 - x_i + y)]\}. \end{aligned} \tag{10.16}$$

For both double penalization dBP_{AC}–OWA $min\{[1 \wedge (1 - k_{Di} + y)]\} \leq f((x_1, y), \ldots, (x_n, y)) \leq max\{[1 \wedge (1 - k_{Di} + y)]\}$.

Theorem 4. (Idempotency). Assume f is the BP_{AC}–IOWA operator; if $[1 \wedge (1 - x_i + y)] = a$ for all i, then

$$f((u_1, ac_1, y), \ldots, (u_n, ac_n, y)) = a. \tag{10.17}$$

For both double penalization dBP_{AC}–OWA operator if $[1 \wedge (1 - k_{Di} + y)] = a$ for all i, then $f((k_{D1}, y), \ldots, (k_{Dn}, y)) = a$.

Theorem 5. (non-negativity). Assume f is the BP_{AC}–OWA operator then:

$$f((ac_1, y), \ldots, (ac_n, y)) \geq 0. \tag{10.18}$$

For double penalization dBP_{AC}–OWA operator $f((k_{D1}, y), \ldots, (k_{Dn}, y)) \geq 0$ operator $f((k_{D1}, y), \ldots, (k_{Dn}, y)) \geq 0$.

Theorem 6. (Reflexivity). Assume f is the BP_{AC}–OWA operator then:

$$f((ac_1, y), \ldots, (ac_n, y)) = 0. \tag{10.19}$$

For double penalization dBP_{AC}–OWA operator $f((k_{D1}, y), \ldots, (k_{Dn}, y)) = 0$.

In addition, if $q = 0$, then BP_{AC}–OWA and dBP_{AC}–OWA, they follow that:

$$BP_{AC} - IOWA^{r,0}\left((ac_1, y), \ldots, (ac_n, y)\right) = \left(\frac{1}{n}\sum_{k=1}^{n} ac_i^r \left(\frac{1}{n-1}\sum_{\substack{j=1 \\ j \neq k}}^{n} ac_j^0\right)\right)^{\frac{1}{r+0}} = \left(\frac{1}{n}\sum_{k=1}^{n} ac_i^r\right)^{\frac{1}{r}}. \tag{10.20}$$

$$dBP_{AC}^{\ r,0}\left((k_{D1}, y), \ldots, (k_{Dn}, y)\right) = \left(\frac{1}{n}\sum_{k=1}^{n} ac_i^r \left(\frac{1}{n-1}\sum_{\substack{j=1 \\ j \neq k}}^{n} ac_j^0\right)\right)^{\frac{1}{r+0}} = \left(\frac{1}{n}\sum_{k=1}^{n} k_{Di}^{\ r}\right)^{\frac{1}{r}}. \tag{10.21}$$

If $r = 2$ and $q = 0$, then BP_{AC}–OWA and dBP_{AC}–OWA reduces to square distance mean:

$$BP_{AC}^{\ 2,0}\left((ac_1, y), \ldots, (ac_n, y)\right) = \left(\frac{1}{n}\sum_{k=1}^{n} ac_i^2\right)^{\frac{1}{2}}. \tag{10.22}$$

$$dBP_{AC}^{\ 2,0}\left((k_{D1}, y), \ldots, (k_{Dn}, y)\right) = \left(\frac{1}{n}\sum_{k=1}^{n} k_{Di}^{\ 2}\right)^{\frac{1}{2}}. \tag{10.23}$$

If $r = 1$ and $q = 0$, then BP_{AC}–OWA and dBP_{AC}–OWA reduces to the usual distance average:

$$BP_{AC}^{\ 1,0}\left((ac_1, y), \ldots, (ac_n, y)\right) = \frac{1}{n}\sum_{k=1}^{n} ac_i. \tag{10.24}$$

$$dBP_{AC}^{\ 1,0}\left((k_{D1}, y), \ldots, (k_{Dn}, y)\right) = \frac{1}{n}\sum_{k=1}^{n} k_{Di}. \tag{10.25}$$

If $r \to +\infty$ and $q = 0$, then BP_{AC}–OWA and dBP_{AC}–OWA reduces to the distance max operator

$$\lim_{r \to +\infty} BP_{AC}^{\ r,0}\left((ac_1, y), \ldots, (ac_n, y)\right) = max\{ac_i\}. \tag{10.26}$$

$$\lim_{r \to +\infty} dBP_{AC}^{\ r,0}\left((k_{D1}, y), \ldots, (k_{Dn}, y)\right) = max\{k_{Di}\}. \tag{10.27}$$

If $r \to 0$ and $q = 0$, then BP_{AC}–OWA and dBP_{AC}–OWA reduces to the geometric distance mean:

$$\lim_{r \to +\infty} BP_{AC}{}^{r,0}\left(\left(ac_1, y\right), \ldots, \left(ac_n, y\right)\right) = \left(\prod_{i=1}^{n} ac_i\right)^{\frac{1}{n}}. \tag{10.28}$$

$$\lim_{r \to +\infty} dBP_{AC}{}^{r,0}\left(\left(k_{D1}, y\right), \ldots, \left(k_{Dn}, y\right)\right) = \left(\prod_{i=1}^{n} k_{Di}\right)^{\frac{1}{n}}. \tag{10.29}$$

If $r = q = 1$, then BP_{AC}–OWA and dBP_{AC}–OWA reduces to the following distance expression:

$$BP_{AC}{}^{1,1}\left(\left(ac_1, y\right), \ldots, \left(ac_n, y\right)\right) = \left(\frac{1}{n(1-n)} \sum_{\substack{k,j=1 \\ j \neq k}}^{n} ac_i ac_j\right)^{\frac{1}{2}}. \tag{10.30}$$

$$dBP_{AC}{}^{1,1}\left(\left(k_{D1}, y\right), \ldots, \left(k_{Dn}, y\right)\right) = \left(\frac{1}{n(1-n)} \sum_{\substack{k,j=1 \\ j \neq k}}^{n} k_{Di} k_{Dj}\right)^{\frac{1}{2}}. \tag{10.31}$$

10.5 PENALTY-BASED PB_{AC}-OWA IN COMBINATION WITH GENERALIZED AND QUASI-ARITHMETIC AGGREGATION OPERATORS

Furthermore, generalized and quasi-arithmetic means can also be employed to extend the BP_{AC}–OWA, dBP_{AC}–OWA operators (Fodor et al., 1995; Yager, 2004). Thus, the Bonferroni penalty generalized ordered weighted averaging adequacy coefficient (BP_{AC}-GOWA) operator, presents particular cases from their general formulation. These new formulations are defined as follows:

Definition 6. The BP_{AC}-GOWA operator is a generalized mean type of continuous aggregation operator that is defined as follows:

$$BP_{AC} - GOWA\left(\left(\mathbf{x}, u, y\right)\right) = \left(\frac{1}{n} \sum_{i} \left(pb_i^r\right)^{\lambda} \left(\frac{1}{n-1} \sum_{\substack{j=1 \\ j \neq i}}^{n} \left(pb_j^q\right)^{\delta}\right)^{\frac{1}{r+q}}\right)^{\frac{1}{\lambda}}, \tag{10.32}$$

$$dBP_{AC} - GOWA((x,u,y)) = \left(\left(\frac{1}{n}\sum_i (dpb_i^r)^\lambda \left(\frac{1}{n-1}\sum_{\substack{j=1\\j\neq i}}^{n} (dpb_j^q)^\delta\right)\right)^{\frac{1}{r+q}}\right)^{\frac{1}{\lambda}}, \quad (10.33)$$

where pb_i is the pAC_i and $dpAC_i$ value of the (BP$_{AC}$-GOWA and dBP$_{AC}$-GOWA pair $\langle x, y\rangle$ having the jth largest x_i, λ and δ are the parameters such that $\lambda \in (-\infty, \infty)$ and $\left(\frac{1}{n-1}\Sigma_{\substack{j=1\\j\neq i}}^{n} (pb_j^q)^\delta\right)$ can be expressed as $P_{AC}OWA_W(V^i)$ where (V^i) is the vector of all $p[1\wedge(1-x_j+y)]$ except $p[1\wedge(1-x_i+y)]$ and $\left(\frac{1}{n-1}\Sigma_{\substack{j=1\\j\neq i}}^{n} (dpb_j^q)^\delta\right)$ can be expressed as $dP_{AC}OWA_W(V^i)$ where (V^i) is the vector of all $dp[1\wedge(1-k_{Dj}+y)]$ except $dp[1\wedge(1-k_{Di}+y)]$ and for both w being an $n-1$ vector W_i associated with α_i whose components w_{ij} are the OWA weights.

Likewise, PB$_{AC}$-IOWA and dPB$_{AC}$-IOWA operators can be extended using quasi-arithmetic means. This results in the quasi-arithmetic QPB$_{AC}$-OWA and QdPB$_{AC}$-OWA, operators, which is defined as follows:

Definition 7. The QPB$_{AC}$-OWA operator is a generalized mean type of continuous aggregation operator that can be defined as follows:

$$BP_{AC} - QOWA((x,y)) = g^{-1}(h^{-1})\left(\frac{1}{n}\sum_i g(pb_i^r)\left(\frac{1}{n-1}\sum_{\substack{j=1\\j\neq i}}^{n} h(pb_j^q)\right)\right)^{\frac{1}{r+q}}, \quad (10.34)$$

$$dBP_{AC} - QOWA((x,y)) = g^{-1}(h^{-1})\left(\frac{1}{n}\sum_i g(dpb_i^r)\left(\frac{1}{n-1}\sum_{\substack{j=1\\j\neq i}}^{n} h(dpb_j^q)\right)\right)^{\frac{1}{r+q}}, \quad (10.35)$$

where pb_i is the pAC_i and $dpAC_i$ value of the PB$_{AC}$-QOWA and dPB$_{AC}$-QOWA pair (x, y) having the jth largest x_i, g and h are strictly continuous monotonic functions and $\left(\frac{1}{n-1}\Sigma_{\substack{j=1\\j\neq i}}^{n} phb_j^q\right)$ can be expressed as $BP_{AC}QOWA_W(V^i)$ where (V^i) is the vector of all $p[1\wedge(1-x_j+y)]$ except $p[1\wedge(1-x_i+y)]$ and

$\left(\frac{1}{n-1}\sum_{\substack{j=1\\j\neq i}}^{n} dphb_j^q\right)$ can be expressed as $dP_{AC}QOWA_W\left(V^i\right)$ where $\left(V^i\right)$ is the vector of all $dp\left[1\wedge\left(1-k_{Dj}+y\right)\right]$ except $dp\left[1\wedge\left(1-k_{Di}+y\right)\right]$ and w being an $n-1$ vector W_i associated with α_i whose components w_{ij} are the OWA weights.

Additionally, Table 10.1 shows particular cases that it is important to mention.

TABLE 10.1
Particular Cases of the BP_{AC}-GOWA and BP_{AC}-QOWA Operators

Particular Cases	BP_{AC}-GOWA	BP_{AC}-QOWA
BP_{AC}-OWA	$\lambda=1,\ \delta=1$	$g=\left(pb_i^r\right),\ h=\left(pb_j^q\right)$
Harmonic BP_{AC}-OWA	$\lambda=-1,\ \delta=-1$	$g=\left(pb_i^r\right)^{-1},\ h=\left(pb_j^q\right)^{-1}$
Quadratic BP_{AC}-OWA	$\lambda=2,\ \delta=2$	$g=\left(pb_i^r\right)^{2},\ h=\left(pb_j^q\right)^{2}$
Cubic BP_{AC}-OWA	$\lambda=3,\ \delta=3$	$g=\left(pb_i^r\right)^{3},\ h=\left(pb_j^q\right)^{3}$
Geometric BP_{AC}-OWA	$\lambda\to 0,\ \delta\to 0$	$g=\left(pb_i^r\right)^{0},\ h=\left(\mathrm{p}\right)^{0}$
BP_{AC}-Max	$\lambda=\infty,\ \delta=\infty$	$g=\left(pb_i^r\right)^{\infty},\ h=\left(pb_j^q\right)^{\infty}$
BP_{AC}-Min	$\lambda=-\infty,\ \delta=-\infty$	$g=\left(pb_i^r\right)^{-\infty},\ h=\left(pb_j^q\right)^{-\infty}$

10.6 MATHEMATICAL EXAMPLE

In a decision-making problem involving disagreements, the following numerical example is presented (see Figure 10.1)

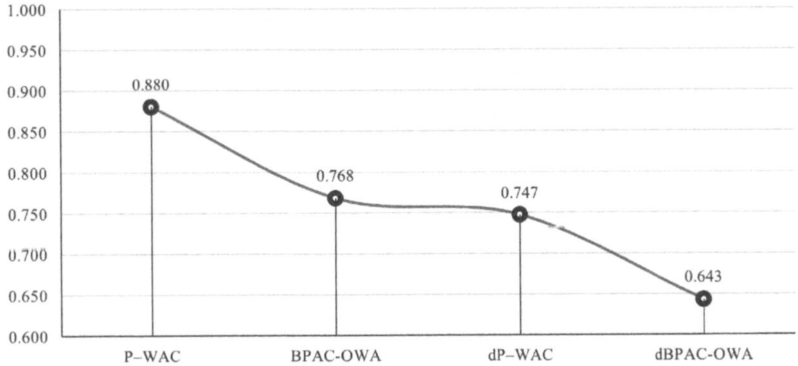

FIGURE 10.1 Graphical representation penalty and double penalty with adequacy coefficient distance.

Example 1

The arguments are $x_i = (0.4,\ 0.5,\ 0.8,\ 1)$

Weighted vector $w_j = (0.8,\ 0.9, 0.7, 06)$

Here, we are considering that $w_i = \left(w_j / \sum_{j=1}^{n} w_j\right)$, then $v_i = (0.27,\ 0.30,\ 0.23,\ 0.20)$

Considering y as mean (merged value), $y = (0.5,\ 0.3,\ 1,\ 0.6)$, then $y = \frac{(0.5+0.3+1+0.6)}{4}$, therefore y = 0.6.

Finally, the ideal argument for each real. Calculating penalty for both:

$$\boldsymbol{P-WAC} = \big(0.27[1\wedge(1-0.5+0.6)]+0.30[1\wedge(1-0.3+0.6)] +0.23[1\wedge(1-1+0.6)]+0.20[1\wedge(1-0.6+0.6)]\big)=\mathbf{0.907}$$

$$\boldsymbol{dP-WAC} = \Big(0.27\big[1\wedge\big(1-(|0.4-0.5|)+0.6\big)\big]+0.30\big[1\wedge\big(1-(1\wedge(1+0.5-0.3))+0.6\big)\big] +0.23\big[1\wedge\big(1-(|0.8-1|)+0.6\big)\big]+0.20\big[1\wedge\big(1-(1\wedge(1+1-0.6))+0.6\big)\big]\Big)=\mathbf{0.320}$$

Example 2

We have assumed that BP_{AC}–OWA and dBP_{AC}–OWA is given by $y_i = (0.5,\ 0.3,\ 1,\ 0.6)$ with weighted vector $w_j = (0.7,\ 0.6, 0.9, 0.8)$. Here, we are considering that $w_i = \left(w_j / \sum_{j=1}^{n} w_j\right)$, then $v_i = (0.23,\ 0.20, 0.30, 0.27)$. Considering y as mean (merged value), then $y = \frac{(0.5+0.3+1+0.6)}{4}$, therefore $y = 0.6$. Finally, the ideal argument for each real argument is $x_i = (0.4,\ 0.5,\ 0.8,\ 1)$ and we take $r = q = 1$.

Now, we calculate penalty for BP_{AC}–OWA:

$$BP_{AC}-OWA_{v_1}\left(V^1\right)=0.23\times\big([1\wedge(1-0.3+0.6)]+[1\wedge(1-1+0.6)]+[1\wedge(1-0.6+0.6)]\big) = 0.607.$$

Following the same procedure, we obtain $BP_{AC}-OWA_{v_2}\left(V^2\right)=0.520$; $BP_{AC}-OWA_{v_3}\left(V^3\right)=0.900$; $BP_{AC}-OWA_{v_4}\left(V^4\right)=0.693$. therefore:

$$BP_{AC}-OWA = \bigg(\frac{1}{4}\times\Big(\big([1\wedge(1-0.5+0.6)]\times0.607\big)+\Big(\big([1\wedge(1-0.3+0.6)]\times0.520\big) +\Big(\big([1\wedge(1-1+0.6)]\times0.900\big)+\Big(\big([1\wedge(1-0.6+0.6)]\times0.693\big)\Big)\Big)\bigg)^{0.5}=0.768.$$

Now, we calculate penalty for dBP–OWA$_{AC}$:

$$dBP_{AC} - OWA_{v_1}\left(V^1\right)$$
$$= 0.23 \times \left(\left[1 \wedge \left(1 - \left(1 \wedge (1 - 0.5 + 0.3)\right) + 0.6\right)\right] + \left[1 \wedge \left(1 - \left(1 \wedge (1 - 0.8 + 1)\right) + 0.6\right)\right]\right.$$
$$\left. + \left[1 \wedge \left(1 - \left(1 \wedge (1 - 1 + 0.6)\right) + 0.6\right)\right]\right) = 0.560.$$

Following the same procedure, we obtain $dBP_{AC} - OWA_{v_2}\left(V^2\right) = 0.440$; $dBP_{AC} - OWA_{v_3}\left(V^3\right) = 0.720$; $dBP_{AC} - OWA_{v_4}\left(V^4\right) = 0.533$. Therefore:

$$dBP_{AC} - OWA$$
$$= \left(\frac{1}{4} \times \left(\left(\left[1 \wedge \left(1 - \left(1 \wedge (1 - 0.4 + 0.5)\right) + 0.6\right)\right] \times 0.560\right)\right.\right.$$
$$+ \left(\left[1 \wedge \left(1 - \left(1 \wedge (1 - 0.5 + 0.3)\right) + 0.6\right)\right] \times 0.440\right)$$
$$+ \left(\left[1 \wedge \left(1 - \left(1 \wedge (1 + 0.8 - 1)\right) + 0.6\right)\right] \times 0.720\right)$$
$$\left.\left. + \left(\left[1 \wedge \left(1 - \left(1 \wedge (1 - 1 + 0.6)\right) + 0.6\right)\right] \times 0.533\right)\right)\right)^{0.5} = 0.643.$$

There are notable differences in the mathematical example results between $P - WAC$ dP–WAC, the difference is 0.587. The AC is the main factor, imposing a more significant penalty the closer it is to zero. When comparing ideal and real arguments in $dP - WAC$, the ability to select a distance operation that leverages the features of distance measures within the penalty neutralization framework is a significant advantage. The double penalty and comparison of x_i and y_i arguments enable decision makers to prioritize comparison and penalties according to their needs. Furthermore, as the number of restrictions and penalties increases, the decision-making process becomes more intricate due to the need to consider multiple factors for compliance. This proposal offers a solution by allowing for the consideration of additional restrictions under specific conditions.

10.7 CONCLUSIONS

The research presents a penalty function and an aggregation extension combined with the Bonferroni mean and penalty functions. This considers disagreement or dissimilarity in expert opinions. First, there is a new heterogenous function based on penalties that uses the P–WAC and dP–WAC (double penalty) ACs. These are characterized by inputs that are symmetric with penalty-saving weights, and the penalty is neutralized y is superior to the x_i. Second, new aggregation operators are proposed that combine the heterogenous penalty functions proposed with the Bonferroni OWA operator. These are called Bonferroni (double) penalty adequacy coefficient ordered by weighted mean to BP_{AC}–OWA and dBP_{AC}–OWA. These operators facilitate the integration of conflicting and constrained expert information into a unified formula, thereby enabling simultaneous comparison and interrelation of the data. It is also important to note that the operator that includes double penalty focuses on comparing ideal situations, allowing complex and highly uncertain scenarios to be generated with a wide range of considerations when evaluating the information. Finally, the generalized mean and the quasi-arithmetic mean are used to extend the operators, and different mathematical theorems applicable to these formulations are presented. Consequently, these approaches seek to address more complex aggregations, characterized by decision problems where information asymmetry and heterogeneity are highly significant. Finally, in terms of future lines of research, there is the application of these operators in other fields of research and the development of other types of operators for more complex problems, such as group decision-making by consensus (Kacprzyk et al., 2021), the use of logarithmic operators (Alfaro-García et al., 2018), prioritized operators (Perez-Arellano et al., 2020), and probabilistic operators (Merigó, 2012).

REFERENCES

Alfaro-Garcia, V. G., Blanco-Mesa, F., León-Castro, E., & Merigo, J. M. (2022). Bonferroni weighted logarithmic averaging distance operator applied to investment selection decision making. *Mathematics*, *10*(12), 2100. https://doi.org/10.3390/math10122100

Alfaro-García, V. G., Merigó, J. M., Gil-Lafuente, A. M., & Kacprzyk, J. (2018). Logarithmic aggregation operators and distance measures. *International Journal of Intelligent Systems*, *33*(7), 1488–1506. https://doi.org/10.1002/int.21988

Ali, J., & Rasool, W. (2024). Interval-valued q-rung orthopair fuzzy Aczel–Alsina operations-based Bonferroni mean aggregation operators and their applications. *Computational and Applied Mathematics*, *43*(1), 1–53.

Baez-Palencia, D., Olazabal-Lugo, M., & Romero-Muñoz, J. (2019). Business decision making through the ordered weighted average. *Inquietud Empresarial*, *19*(2), 11–23. https://doi.org/10.19053/01211048.9843

Beliakov, G., Bustince Sola, H., & Calvo Sánchez, T. (2016). Review of aggregation functions. In G. Beliakov, H. Bustince Sola, & T. Calvo, (Eds.), *A Practical Guide to Averaging Functions. Studies in Fuzziness and Soft Computing* (Vol. 329, pp. 1–53). Springer, Cham. https://doi.org/10.1007/978-3-319-24753-3_1

Beliakov, G., Calvo, T., & James, S. (2011). On penalty-based aggregation functions and consensus. In E. Herrera-Viedma, J. L. García-Lapresta, J. Kacprzyk, M. Fedrizzi, H. Nurmi, & S. Zadrożny (Eds.), *Studies in Fuzziness and Soft Computing* (Vol. 267, pp. 23–40). Springer, Berlin, Heidelberg. https://doi.org/10.1007/978-3-642-20533-0_2

Beliakov, G., Pradera, A., & Calvo, T. (2007). *Aggregation Functions: A Guide for Practitioners* (Vol. 221). Springer Berlin Heidelberg. https://doi.org/10.1007/978-3-540-73721-6

Blanco-Mesa, Fabio, León-Castro, E., & Merigó, J. M. (2019). A bibliometric analysis of aggregation operators. *Applied Soft Computing*, *81*, 105488. https://doi.org/10.1016/j.asoc.2019.105488

Blanco-Mesa, F., León-Castro, E., & Merigó, J. M. (2020). Covariances with OWA operators and Bonferroni means. *Soft Computing*, *24*(19), 14999–15014. https://doi.org/10.1007/s00500-020-04852-5

Blanco-Mesa, F., Merigó, José M., & Kacprzyk, J. (2016). Bonferroni means with distance measures and the adequacy coefficient in entrepreneurial group theory. *Knowledge-Based Systems*, *111*(1), 217–227. https://doi.org/10.1016/j.knosys.2016.08.016

Blanco-Mesa, F., Rivera-Rubiano, J., Patiño-Hernandez, X., & Martinez-Montaña, M. (2019). The importance of enterprise risk management in large companies in Colombia. *Technological and Economic Development of Economy*, *25*(4). https://doi.org/10.3846/tede.2019.9380

Bonferroni, C. (1950). Sulle medie multiple di potenze. *Bollettino dell'Unione Matematica Italiana*, *5*(3-4), 267–270.

Calvo, T., & Beliakov, G. (2010). Aggregation functions based on penalties. *Fuzzy Sets and Systems*, *161*(10), 1420–1436. https://doi.org/10.1016/J.FSS.2009.05.012

Espinoza-Audelo, L. F., Olazabal-Lugo, M., Blanco-Mesa, F., León-Castro, E., & Alfaro-Garcia, V. (2020). Bonferroni probabilistic ordered weighted averaging operators applied to agricultural commodities' price analysis. *Mathematics*, *8*(8), 1350. https://doi.org/10.3390/math8081350

Fodor, J., Marichal, J. L., & Roubens, M. (1995). Characterization of the ordered weighted averaging operators. *IEEE Transactions on Fuzzy Systems*, *3*(2), 236–240. https://doi.org/10.1109/91.388176

Janani, K., Mohanrasu, S. S., Lim, C. P., Manavalan, B., & Rakkiyappan, R. (2023). Ensemble feature selection using Bonferroni, OWA and Induced OWA aggregation operators. *Applied Soft Computing*, *143*, 110431. https://doi.org/10.1016/J.ASOC.2023.110431

Kacprzyk, J., Zadrozny, S., Nurmi, H., & Bozhenyuk, A. (2021). Towards innovation focused fuzzy decision making by consensus. *IEEE International Conference on Fuzzy Systems, 2021-July*. https://doi.org/10.1109/FUZZ45933.2021.9494531

Kaufmann, A., & Gil-Aluja, J. (1986). Introducción de la teoría de los subconjuntos borrosos a la gestión de las empresas. Milladoiro.

Kaufmann, A., & Gil-Aluja, J. (1987). *Técnicas operativas de gestión para el tratamiento de la incertidumbre*. Hispano Europea. http://dialnet.unirioja.es/servlet/libro?codigo=219261

Leon-Castro, E., Blanco-Mesa, F., & Merigó, J. M. (2020). Bonferroni means with the induced OWAWA operator. *Fuzzy Economic Review*, *25*(1). https://doi.org/10.25102/fer.2020.01.02

Li, Y., & Wang, Q. (2019). Research on hotel supply chain risk assessment with dual generalized triangular fuzzy Bonferroni mean operators. *Journal of Intelligent & Fuzzy Systems*, *37*(2), 1953–1965. https://doi.org/10.3233/JIFS-179257

Liang, D., Darko, A. P., & Xu, Z. (2019). Pythagorean fuzzy partitioned geometric Bonferroni mean and its application to multi-criteria group decision making with grey relational analysis. *International Journal of Fuzzy Systems*, *21*(1), 115–128. https://doi.org/10.1007/s40815-018-0544-x

Liu, W., Chang, J., & He, X. (2023). Intuitionistic fuzzy normalized weighted geometric Bonferroni means with reducibility and boundedness and application in decision making. *Journal of Intelligent & Fuzzy Systems*, *45*(5), 7587–7601. https://doi.org/10.3233/JIFS-231678

Liu, P., & Liu, W. (2021). Dual generalized Bonferroni mean operators based on 2-dimensional uncertain linguistic information and their applications in multi-attribute decision making. *Artificial Intelligence Review*, *54*(1), 491–517. https://doi.org/10.1007/s10462-020-09857-y

Merigó, J. M. (2012). The probabilistic weighted average and its application in multiperson decision making. *International Journal of Intelligent Systems*, *27*(5), 457–476. https://doi.org/10.1002/int.21531

Perez-Arellano, L. A., Blanco-Mesa, F., Leon-Castro, E., & Alfaro-Garcia, V. (2020). Bonferroni prioritized aggregation operators applied to government transparency. *Mathematics*, *9*(1), 24. https://doi.org/10.3390/math9010024

Ren, Y., Yuan, X., & Huang, L. (2022). Q-rung hesitant triangular fuzzy BM operator and its application in multiple criteria decision making. *Systems Engineering and Electronics, 44*(11), 181–191.

Xia, M., Xu, Z., & Zhu, B. (2012). Generalized intuitionistic fuzzy Bonferroni means. *International Journal of Intelligent Systems, 27*(1), 23–47. https://doi.org/10.1002/int.20515

Yager, R. R. (1988). On ordered weighted averaging aggregation operators in multicriteria decision-making. *IEEE Transactions on Systems, Man, and Cybernetics, 18*(1), 183–190. https://doi.org/10.1109/21.87068

Yager, R. R. (2004). Generalized OWA aggregation operators. *Fuzzy Optimization and Decision Making, 3*(1), 93–107. https://doi.org/10.1023/B:FODM.0000013074.68765.97

Yager, R. R. (2009). On generalized Bonferroni mean operators for multi-criteria aggregation. *International Journal of Approximate Reasoning, 50*(8), 1279–1286. https://doi.org/10.1016/j.ijar.2009.06.004

Yang, Y., Hua, L., Jie, M., & Shi, B. (2024). Large-scale satisfaction rating-driven selection of new energy vehicles: A basic uncertain linguistic information Bonferroni mean-based MCGDM approach considering criteria interaction. *Sustainability, 16*(16), 6737. https://doi.org/10.3390/su16166737

11 A Novel Category of Fuzzy β-Compact Spaces

Darshana B. Likhada

11.1 INTRODUCTION

Topology has vast application in many fields like, decision-making problem (Abu-Gdairi et al., 2022), medical (El-Bably and Abo-Tabl, 2021), prediction of lung cancer and heart failure (El-Bably et al., 2021; El-Bably and Abo-Tabl, 2021), biochemistry (Stadler and Stadler, 2022), etc. Thus, the main aim of this study is to contribute into applications of topology by significant development in the theory of topological spaces, especially fuzzy topological spaces.

This chapter is divided into four sections. Motivation of this work and history part is contained in introduction Section 11.1. The root of theory of fuzzy sets begins in 1965, and so beginning with the definition of fuzzy sets, followed by an overview of general topology, fuzzy topological spaces, and other necessary preliminaries, the groundwork required for this work is presented in Section 11.2. Finally, Section 11.3 is dedicated to the new definitions of this novel category of fuzzy β-compact spaces, its interrelations as well as relations with existing fuzzy compact spaces from Hanafy (2004) is established. Some counter examples are also presented to strengthen the relations. Section 11.4 contains the conclusion of this work by presenting the hierarchical diagram.

The fuzzy sets are a specific kind of sets. In the area of set theory, fuzzy mathematics stands different from conventional mathematics. The concept of fuzzy sets was introduced by Zadeh (Zadeh, 1965) in 1965, which laid the foundation for the growth of fuzzy mathematics. The topology defined on fuzzy sets named as fuzzy topological spaces. Fuzzy topology, a branch of fuzzy mathematics and a generalization of general topology, was first introduced by C. L. Chang (Chang, 1968) in 1968. Hanafy (Hanafy, 2004) introduced and studied the fuzzy β-compact spaces and other types of fuzzy compact spaces in 2004, and recently the study of special kind of compact spaces SC-compact spaces is introduced by A. Nasef et al. (Nasef et al., 2025) in 2025 and shows that this SC-compact spaces occupies position between semi-compact and C-compact spaces.

Motivated from this development in general topology and fuzzy topology, this chapter introduces and examines the novel category of fuzzy β-compact spaces, namely fuzzy βSC-compact, fuzzy βC-compact, and fuzzy β-almost compact. The interrelations of this compactness and relation with known compactness are examined here. This rigorous development in the theory of fuzzy compactness provides general concept of C-compact spaces and SC-compact spaces of general topology. Also, this ground work opens many doors into the applications of fuzzy compactness too.

DOI: 10.1201/9781003724902-12

11.2 PRELIMINARIES

Given a non-empty set X and the interval $I = [0,1]$, a fuzzy set (Zadeh, 1965) in X is described by a function $\mu : X \to I$, that is an element of I^X. Here $\mu(t)$ is known as the Grade of membership of $t \in X$ with respect to μ. For completeness, we recall the notions of topological space and compact space as established in general topology.

Definition 1. (Munkres, 2000) Given a non-empty set X, a *topology* τ on X is defined as a family of subsets of X that satisfies the conditions below.

1. The empty set φ and X are in τ.
2. For every $\alpha \in \Lambda$, $A_\alpha \in \tau$ implies $\cup A_\alpha \in \tau$.
3. For $i \in \{1, 2, \ldots, n\}$, $A_i \in \tau$ implies $\cap A_i \in \tau$.

A *topological space* is defined as the ordered pair (X, τ) consisting of a set X and together with a topology τ on it.

Definition 2. (Munkres, 2000) A topological space X is called *compact* if every open cover of X gives a finite subcover.

The topology on fuzzy set $\mathfrak{X}$ may be seen as a collection of characteristic functions on $\mathfrak{X}$, because of the one-oneness between the set power set $\mathcal{P}(\mathfrak{X})$ of $\mathfrak{X}$ and the family of all characteristic functions defined on $\mathfrak{X}$. The standard set operations subset, union, intersection, and complementation in $\mathcal{P}(\mathfrak{X})$ is replaced by function operations $\leq$, $\vee$, $\wedge$ and complement of function as defined below.

Definition 3. (Zadeh, 1965) For each $\mu_1, \mu_2 \in I^{\mathfrak{X}}$ and $t \in \mathfrak{X}$,

1. $(\mu_1 \wedge \mu_2)(t) = \min\{\mu_1(t), \mu_2(t)\}$.
2. $(\mu_1 \vee \mu_2)(t) = \max\{\mu_1(t), \mu_2(t)\}$.
3. $(\mu_1)^c(t) = 1 - \mu_1(t)$.
4. If $f : \mathfrak{X} \to \mathfrak{Y}$, $\mu \in I^{\mathfrak{X}}$ and $\nu \in I^{\mathfrak{Y}}$, then $f(\mu)$ is a fuzzy set in $\mathfrak{Y}$ with $f(\mu)(s) = \sup\{\mu(t) : t \in f^{-1}(s)\}$ whenever $f^{-1}(s) \neq \varphi$, and $f(\mu)(s) = 0$ whenever $f^{-1}(s) = \varphi$. Also, given a fuzzy set ν in $\mathfrak{Y}$, the pre-image $f^{-1}(\nu)$ is a fuzzy set in $\mathfrak{X}$ such that

$$f^{-1}(\nu)(s) = \nu(f(t)), \; (t \in \mathfrak{X}).$$

Similar to the Definition 1 of topology, the following defines a fuzzy topology.

Definition 4. (Chang, 1968) On a non-empty set $\mathfrak{X}$, a *fuzzy topology* $\mathfrak{T}$ is defined as a collection of fuzzy sets ($\mathfrak{T} \subseteq I^{\mathfrak{X}}$) that meets the conditions below.

1. The characteristic functions $\overline{0}$ and $\overline{1}$ are members of $\mathfrak{T}$,
2. If $\mu_i \in \mathfrak{T}$ for every $i \in \mathcal{I}$, then $\vee\{\mu_i : i \in \mathcal{I}\} \in \mathfrak{T}$,
3. If $\mu, \nu \in \mathfrak{T}$, then $\mu \wedge \nu \in \mathfrak{T}$.

The pair $(\mathfrak{X}, \mathfrak{T})$ is known as a *fuzzy topological space*. The elements of $\mathfrak{T}$ are called *open fuzzy sets*, and if the complement of $\mu \in I^{\mathfrak{X}}$ is open fuzzy set then μ is called *closed fuzzy set*.

Throughout this chapter, unless it is specified we consider $\mathfrak{X}$ as fuzzy topological space given with a fuzzy topology $\mathfrak{T}$ on it. In short, we call $\mathfrak{X}$ a fuzzy space. A fuzzy point (Pu and Liu, 1980) $t_\alpha \in \mathfrak{X}$ refers to a fuzzy set whose support is $t \in \mathfrak{X}$ and whose value is $\alpha \in (0,1]$.

Definition 5. A subset S of a non-empty fuzzy space $\mathfrak{X}$ is defined as a

1. semi-open fuzzy set (Azad, 1981) if $S \le Cl(Int(S))$;
2. preopen fuzzy set (Singal and Prakash, 1991) if $S \le Int(Cl(S))$;
3. β-open fuzzy set (Fath alla, 1984) if $S \le Cl(Int(Cl(S)))$;
4. β-closed fuzzy set (Fath alla, 1984) if $\mathfrak{X} \setminus S$ is β-open fuzzy set, i.e., $Int(Cl(Int(S))) \le S$.

It is immediate from the above definition that, each semi-open and preopen fuzzy sets are β-open fuzzy set.

Definition 6. (Hanafy, 2004) Let S be a subset of fuzzy space $\mathfrak{X}$. Then,

1. The intersection of all fuzzy β-closed sets containing S is called fuzzy β-*closure* of S. It is denoted by $\beta Cl(S)$. Here,

$$\beta Cl(S) = \wedge\{C : S \le C,\ C \text{ is } \beta - closed\}.$$

2. The union of all fuzzy β-open sets contained in S is called fuzzy β-*interior* of S. It is denoted by $\beta Int(S)$. Here,

$$\beta Int(S) = \vee\{O : O \le S,\ O \text{ is } \beta - open\}.$$

It is observed in Abu-Gdairi et al. (2023), Hanafy (2004), and (Singh, 2024) that, the fuzzy β-interior and the fuzzy β-closure of a set follows the following standard properties analogous to the properties given in Munkres (2000) from general topology.

In a fuzzy topological space $\mathfrak{X}$, the subsets S and T of $\mathfrak{X}$ satisfy the following properties.

1. $S \leq \beta Cl(S)$
2. $\beta Cl(S)$ is always β-closed set
3. $\beta Cl(\overline{0}_{\mathfrak{X}}) = \overline{0}_{\mathfrak{X}}$
4. $\beta Cl(\beta Cl(S)) \leq \beta Cl(S)$
5. $\beta Cl(S \vee T) = \beta Cl(S) \vee \beta Cl(T)$
6. If S is β-closed set, then $\beta Cl(S) = S$
7. $\beta Int(S) \leq S$
8. $\beta Int(S)$ is always β-open set
9. $\beta Int(\overline{0}_{\mathfrak{X}}) = \overline{0}_{\mathfrak{X}}$ and $\beta Int(\overline{1}_{\mathfrak{X}}) = \overline{1}_{\mathfrak{X}}$
10. $\beta Int(\beta Int(S)) \leq \beta Int(S)$
11. $\beta Int(S \wedge T) = \beta Int(S) \wedge \beta Int(T)$
12. If S is β-open set, then $\beta Int(S) = S$

In order to define compact spaces in fuzzy topological spaces, the concept of cover and subcover of a set is introduced by Ganguly and Saha (Ganguly and Saha, 1990) in 1990. Their definition of cover and subcover in fuzzy settings is as follows.

Definition 7. (Ganguly and Saha, 1990) Let $\mathfrak{X}$ be a fuzzy topological space and A be a fuzzy set in $\mathfrak{X}$. A collection $\{A_i : i \in I\}$ of fuzzy sets of $\mathfrak{X}$ is called a *fuzzy cover* of A if $A \leq \vee_{i \in I} A_i$. A fuzzy cover is said to be open if each A_i is open fuzzy set. If there exists a subset I_1 of I such that $A \leq \vee_{i \in I_1} A_i$, then $\{A_i : i \in I_1\}$ is called a *fuzzy subcover* of fuzzy cover $\{A_i : i \in I\}$.

Following the concept of open covers in fuzzy topology and the availability of different fuzzy open sets like, preopen, semi-open, and β-open, researchers have formulated multiple notions of fuzzy compactness based on the chosen type of fuzzy open sets. The definitions of all these various fuzzy compact spaces and types of compact spaces are put here to serve the primary idea, which may be helpful in the next section.

Definition 8. A fuzzy space $\mathfrak{X}$ is

1. *strongly compact* (Nanda, 1991) whenever every preopen cover of $\mathfrak{X}$ admits a finite subcover.
2. *semi-compact* (Mashhour et al., 1987) whenever every semi-open cover of $\mathfrak{X}$ admits a finite subcover.
3. *almost compact* (Di Concillio and Gerla, 1984) whenever every open cover of $\mathfrak{X}$ admits a finite subcollection whose closures cover $\mathfrak{X}$.
4. *S-closed* (Coker and Haydar, 1987) whenever every semi-open cover of $\mathfrak{X}$ admits a finite subcollection whose closures cover $\mathfrak{X}$.
5. *s-closed* (Di Maio and Noiri, 1987) whenever every semi-open cover of $\mathfrak{X}$ admits a finite subcollection whose semi-closures cover $\mathfrak{X}$.
6. *P-closed* (Zahran, 1995) whenever every preopen cover of $\mathfrak{X}$ admits a finite subcollection whose pre-closures cover $\mathfrak{X}$.
7. *β-compact* (Balasubramanian, 1997) whenever every β-open cover of $\mathfrak{X}$ admits a finite subcover.

C. Dorsett (Dorsett, 1981) introduced the idea of C-compact spaces in general topology in 1981.

Definition 9. (Dorsett, 1981) In a topological space (X,τ), if every closed subset of X is almost compact with respect to the topology τ, then X is called *C-compact.*

In 2025, recent study has established a new type of compact spaces which takes place between the semi-compact spaces and C-compact spaces, A. Nasef et al. (Nasef et al., 2025) call this compactness an SC-compactness and confirmed the extension of classical compactness properties by SC-compact spaces. The following is the definition of SC-compact space.

Definition 10. (Nasef et al., 2025) A topological space (X,τ) is called *SC-compact* if each semi-closed subsets of X is s-closed relative to the topology τ. That is, each semi-closed subset must be covered with finitely many semi-closures of semi-open sets that together form a cover of the subset.

11.3 MAIN RESULTS

Motivated from the recent study (Nasef et al., 2025) on SC-compact spaces and its relation to semi-compact and S-closed spaces in general topology, here are the analogous extended concepts in the fuzzy topological set up.

Definition 11. A fuzzy space $\mathfrak{X}$ is *almost* β-*compact* whenever each β-open cover can be reduced to a finite subcover whose β-closures together cover $\mathfrak{X}$.

It is evident from the definition that every fuzzy β-almost compact space is also a fuzzy almost compact as β-closure of a set contains closure of a set. Hence, β-almost compactness is the general version of fuzzy almost compactness. Further, fuzzy β-compactness implies fuzzy β-almost compactness as if there is a finite subcover of every β-open cover, then β-closure of elements of this subcover will be also a cover.

Definition 12. A fuzzy space $\mathfrak{X}$ is said to be βC-*compact* if each β-closed subset of $\mathfrak{X}$ is β-almost compact with respect to the fuzzy topology on $\mathfrak{X}$.

Definition 13. A fuzzy space $\mathfrak{X}$ is said to be βSC-*compact* if each semi-closed subsets of $\mathfrak{X}$ can be covered with finitely many β-closures of β-open sets that together form a cover of the subset.

The notion of C-compactness (Dorsett, 1981) and SC-compactness (Nasef et al., 2025) in general topology is established as mentioned earlier, so it is obvious that one can similarly talk of fuzzy C-compactness and fuzzy SC-compactness in fuzzy topological space. So, now we are ready to compare the fuzzy βC-compactness and fuzzy C-compactness, as well as βSC-compactness and fuzzy SC-compactness.

Remark 1. Fuzzy βSC-compact $\Rightarrow$ fuzzy SC-compact. For that, as semi-open fuzzy sets imply β-open fuzzy sets, the corresponding β-closed sets are necessarily semi-closed. Thus, it is clear that β-closure of a set contains semi-closure of a set.

Theorem 1. (Fuzzy semi-compact $\Rightarrow$ fuzzy SC-compact) Every fuzzy semi-compact space is always a fuzzy SC-compact space.

Proof. Consider a fuzzy semi-compact space $\mathfrak{X}$ with Y as a fuzzy semi-closed subset of $\mathfrak{X}$. Take a fuzzy semi-open cover $\mathcal{A}$ of Y. Since Y is a fuzzy semi-closed set, the complement Y^c is fuzzy semi-open. So, $\mathcal{A}^* = \mathcal{A} \vee Y^c$ gives a semi-open cover for $\mathfrak{X}$. Since $\mathfrak{X}$ being a fuzzy semi-compact space, $\mathcal{A}^*$ admits finite subcover. If this finite subcover contains Y^c then discard it, otherwise keep as it is. Hence, this finite subcover will cover Y, and so semi-closures of elements of this finite subcover will also cover Y. This shows that $\mathfrak{X}$ is fuzzy SC-compact.

Remark 2. It is evident that, fuzzy SC-compact ⇒ fuzzy βC-compact. Indeed, each β-closed subset is semi-closed and semi-closure of semi-open set is same as β-closure of β-open sets.

Theorem 2. (Fuzzy βC-compact ⇒ fuzzy C-compact) Every fuzzy βC-compact space is always fuzzy C-compact space.

Proof. Consider a fuzzy βC-compact space $\mathfrak{X}$ with Y as a fuzzy closed subset of $\mathfrak{X}$. Take a fuzzy open cover $\mathcal{A}$ of Y, then $\mathcal{A}$ is also a β-open cover of Y. Since Y is a fuzzy closed set, the complement Y^c is fuzzy open and so β-open. Now, consider a β-open cover $\mathcal{A}^* = \mathcal{A} \vee Y^c$ of $\mathfrak{X}$. Since $\mathfrak{X}$ given as a fuzzy βC-compact space and $\mathfrak{X}$ itself a β-closed set in $\mathfrak{X}$, $\mathfrak{X}$ is fuzzy β-almost compact. Thus, there exists a finite subcollection $\mathfrak{B}$ of $\mathcal{A}^*$ whose β-closures cover $\mathfrak{X}$. If $\mathfrak{B}$ contains Y^c, then discard it otherwise keep as it is. So, Y is covered by closure of elements of $\mathfrak{B}$ as Y is covered by β-closure of elements of $\mathfrak{B}$. Therefore Y is fuzzy almost compact and that proves $\mathfrak{X}$ is fuzzy C-compact.

Remark 3. By above observations, it is clear that the fuzzy β-almost compactness, fuzzy βC-compactness and fuzzy βSC-compactness is the extension of almost compactness, C-compactness, and SC-compactness, respectively.

The following relation is very evident from the definitions of both the space. But for the sake of completeness, the proof is included here.

Theorem 3. (Fuzzy β-compact ⇒ fuzzy β-almost compact) Every fuzzy β-compact space is fuzzy β-almost compact.

Proof. Consider a fuzzy β-compact space $\mathfrak{X}$ and $\mathcal{A}$ to be β-open cover of $\mathfrak{X}$. As $\mathfrak{X}$ is fuzzy β-compact, there is finite subcollection $\mathfrak{B}$ of $\mathcal{A}$ which covers $\mathfrak{X}$. Then, as set is always contained in its β-closure, clearly β-closure of the all elements of $\mathfrak{B}$ will also cover $\mathfrak{X}$. Thus, $\mathfrak{X}$ is β-almost compact.

Theorem 4. (Fuzzy β-almost compact ⇒ fuzzy strongly compact) Every fuzzy β-almost compact space is fuzzy strongly compact.

Proof. Let $\mathfrak{X}$ be a fuzzy β-almost compact space and $\mathcal{A}$ be a preopen cover of $\mathfrak{X}$. Since preopen sets are β-open, $\mathcal{A}$ will be β-open cover for $\mathfrak{X}$. Also, $\mathfrak{X}$ being a β-almost compact, there is a finite subcollection $\mathfrak{B}$ of $\mathcal{A}$ whose β-closures cover $\mathfrak{X}$. Thus, closure of elements of $\mathfrak{B}$ covers $\mathfrak{X}$. Therefore, $\mathfrak{X}$ is strongly compact.

The following example makes the statement of Theorem 4 much stronger by providing the counter example for the converse of Theorem 4.

Example 1. Consider a non-empty fuzzy set X with $\mu_k(t) = 1 - \frac{1}{k}$ for all $t \in X$ and $k \in \mathbb{N}$. Then the collection $\{\mu_k : k \in \mathbb{N}\}$ is a β-open cover for X and $\vee \beta Cl(\mu_k) = \overline{1}_X$, for $k \geq 3$. This shows that X is β-almost compact but not β-compact.

Theorem 5. (Fuzzy strongly compact ⇒ fuzzy β*C*-compact) Every fuzzy strongly compact space is always fuzzy β*C*-compact space.

Proof. Consider a fuzzy strongly compact space $\mathfrak{X}$ with β-closed subset Y. For any β-open cover $\mathcal{A}$ for Y, $\mathcal{A}^* = \mathcal{A} \vee Y^c$ will be a β-open cover for $\mathfrak{X}$ because Y is β-closed. That is,

$$\mathfrak{X} = \vee_{E \in \mathcal{A}^*} Cl(Int(Cl(E))).$$

Rewriting it,

$$\mathfrak{X} = \vee_{H = Int(Cl(E)),\ E \in \mathcal{A}^*} Cl(H).$$

Call this collection $\mathfrak{H}$ of all preopen sets $H = Int(Cl(E))$, $E \in \mathcal{A}^{\star}$. So, $\mathfrak{H}$ will be a preopen cover for $\mathfrak{X}$. As $\mathfrak{X}$ is fuzzy strongly compact space, the preopen cover $\mathfrak{H}$ admits finite subcover $\mathfrak{H}'$ (say). If $\mathfrak{H}'$ contains Y^c then discard it, otherwise keep $\mathfrak{H}'$ as it is. It is clear that β-closure of finitely many elements of $\mathfrak{H}'$ cover Y. As Y was arbitrary β-closed subset of $\mathfrak{X}$, it shows that $\mathfrak{X}$ is fuzzy βC-compact.

The converse of above statement need not hold. The following is the counter example.

Example 2. If X is any non-empty fuzzy set and $\mu_k(t) = 1 - \frac{1}{k}$ for all $t \in X$ and $k \in \mathbb{N}$. Then each μ_k is β-open fuzzy set and $\{\mu_k : k \in \mathbb{N}\}$ is a β-open cover for X. Also, $\vee \beta Cl(\mu_k) = \overline{1}_X$, for $k \geq 3$ shows that X is β-closed, it is not β-compact. Moreover, this cover $\{\mu_k : k \in \mathbb{N}\}$ is also a preopen cover for X, that does not admit finite (preopen) subcover. So, X cannot be a fuzzy strongly compact.

11.4 CONCLUSION

The compact spaces plays an important role in the theory of topology, with many foundation results which strengthen the theory itself and applications too. Fuzzy topological space has many practical real-life applications in engineering, decision making, design making, treatment of lung cancers, biochemistry, etc. The introduction of fuzzy β-compactness in Balasubramanian (1997) is a milestone in the fuzzy compactness theory, as it has open many doors to extend the theory of compactness and thus the related foundations and applications too. The development in Nasef et al. (2025) about SC-compactness has motivated for this research and showcase the concern about this category of compact spaces in fuzzy topological spaces, and thus the development of fuzzy β-almost compactness, fuzzy βC-compactness, and βSC-compactness is carries out. Also, the interrelations of this new compact spaces and the relation with existing fuzzy compact spaces are established. Some counter examples are also provided in order to fulfil that converse of the stated results are not true.

Putting all the theorems and remarks of previous section together, the following diagram summarizes the work done and gives a clear idea of exact hierarchy of fuzzy compact spaces. Here, F stands for fuzzy.

F β-compact	$\Rightarrow$	F β-almost compact	$\Rightarrow$	F strongly compact		
$\Downarrow$				$\Downarrow$		
F semi-compact	$\Rightarrow$	F SC-compact	$\Rightarrow$	F βC-compact	$\Rightarrow$	F C-compact
		$\Uparrow$				
		F βSC-compact				

It is observed from the given diagram that the position of fuzzy β-almost compactness is between the fuzzy β-compactness and fuzzy strongly compactness. Also, fuzzy βC-compactness lies between fuzzy βSC-compactness and fuzzy C-compactness.

REFERENCES

Abu-Gdairi, R., El-Gayar, M. A., Al-shami, T. M., Nawar, A. S. & El-Bably, M. K. (2022). *Some topological approaches for generalized rough sets and their decision making applications*, Symmetry, 14 (1). https://doi.org/10.3390/sym14010095

Abu-Gdairi, R., Nasef, A. A., El-Gayar, M. A. & El-Bably, M. K. (2023). *On fuzzy point applications of fuzzy topological spaces.* International Journal of Fuzzy Logic and Intelligent Systems, 23(2), 162–172.

Azad, K. K. (1981). *On fuzzy semicontinuity, fuzzy almost continuity, and fuzzy weakly continuity.* Journal of Mathematical Analysis and Applications, 82(1), 14–32. https://doi.org/10.1016/0022-247X(81)90222-5

Balasubramanian, G. (1997). *On fuzzy β-compact spaces and fuzzy β-extremally disconnected spaces* [cybernetics], Kybernetika, 33(3), 271–277.

Chang, C. L. (1968). *Fuzzy topological spaces.* Journal of Mathematical Analysis and Applications, 24(1), 182–190. https://doi.org/10.1016/0022-247X(68)90057-7

Coker, D. & Haydar, A. (1987). *On fuzzy s-closed spaces.* Doga Turkish Journal of Mathematics, 145–152.

Di Concillio, A. & Gerla, G. (1984). *Almost compactness in fuzzy topological spaces.* Fuzzy Sets and Systems, 13, 187–192.

Di Maio, G. & Noiri, T. (1987). *On s-closed spaces.* Indian Journal of Pure and Applied Mathematics, 18(3), 226–233.

Dorsett, C. (1981). *Semi-compactness, semi-separation axioms and product spaces.* Bulletin of the Malaysian Mathematical Sciences Society, 2(4), 21–28.

El-Bably, M. K. & Abo-Tabl, E. A. (2021). *A topological reduction for predicting of a lung cancer disease based on generalized rough sets.* Journal of Intelligent and Fuzzy Systems, 41(2), 3045–3060. https://doi.org/10.3233/JIFS-210167

El-Bably, M. K., Ali, M. I. & Abo-Tabl, E. A. (2021). *New topological approaches to generalized soft rough approximations with medical applications.* Journal of Mathematics. https://doi.org/10.1155/2021/2559495

Fath alla, M. A., *On fuzzy topological spaces*, Ph.D. dissertation, Assuit University, Sohag, Egypt, 1984.

Ganguly, S. & Saha, S. (1990). *A note on compactness in fuzzy setting.* Fuzzy Sets and Systems, 34, 117–124.

Hanafy, I. M. (2004). *Fuzzy β-compactness and fuzzy β-closed spaces.* Turkish Journal of Mathematics, 28(3), 282–293.

Mashhour, A. S., Allam, A. A. & Abd El-Hakeim, K. M. (1987). *On fuzzy semicompact spaces.* Bulletin of the Faculty of Science, Assiut University, 16(1), 277–285.

Munkres, J., Topology, 2nd ed., PHI learning, 2000.

Nanda, S. (1991). *Strongly compact fuzzy topological spaces.* Fuzzy Sets and Systems, 42, 259–262.

Nasef, A. A., El-Bably, M. K. & El-Gayar, M. A. (2025). *On SC-compact spaces and their relations to semi-compact and s-closed spaces.* Advances in Basic and Applied Sciences, 5, 29–37.

Pu, P. M. & Liu, Y. M. (1980). *Fuzzy topology. I. Neighborhood structure of a fuzzy point and moore-smith convergence*, Journal of Mathematical Analysis and Applications, 76(2), 571–599. https://doi.org/10.1016/0022-247X(80)90048-7

Singal, M. K. & Prakash, N. (1991). *Fuzzy preopen sets and fuzzy preseparation axioms.* Fuzzy Sets and Systems, 44, 273–281.

Singh, M. K., Applied Fuzzy Mathematics, Forum for Interdisciplinary Math., Springer Nature Singapore Pte Ltd. 2024.

Stadler, B. M. R. & Stadler, P. F. (2022). *Generalized topological spaces in evolutionary theory and combinatorial chemistry*, Journal of Chemical Information and Modelling, 42(3), 577–585. https://doi.org/10.1021/ci0100898

Zadeh, L. A. (1965). *Fuzzy sets.* Information and Control, 8(3), 338–353. https://doi.org/10.1016/S0019-9958(65)90241-X

Zahran, A. M. (1995). *Strongly compact and P-closed fuzzy topological spaces.* Journal of Fuzzy Mathematics, 3(1), 97–102.

Conclusion

Ernesto León-Castro, Fabio Raul Blanco Mesa, Victor Alfaro-Garcia, and Anna Maria Gil-Lafuente

The eleven chapters gathered in this volume collectively advance a proposition that is both methodological and epistemological: that the most consequential decisions in management, engineering, finance, public policy, and public health are made by people navigating conditions of uncertainty, competing criteria, and heterogeneous information, and that the mathematical and fuzzy frameworks developed within the tradition of multicriteria decision making (MCDM) and aggregation theory constitute some of the most rigorous and practically productive tools available to support human judgment under those conditions. This concluding chapter synthesizes the contributions of the volume and charts the directions along which this area of research can advance.

The volume starts with a presentation of the fundamentals and the construction of mathematical foundations, proceeds through the development of computational decision architectures, demonstrates these tools across diverse applied domains, and finally identifies research frontiers where these methodologies are most urgently needed. This progression reflects the conviction that the field of mathematical and fuzzy decision science advances most productively when theoretical innovation, applied validation, and the mapping of unresolved problems are pursued as complementary activities within a shared intellectual project, one ultimately oriented toward enabling better decisions by the people and institutions that face them.

C.1 THE MATHEMATICAL FOUNDATIONS: AGGREGATION THEORY AND FUZZY TOPOLOGICAL INFRASTRUCTURE

Every applied contribution in this volume depends, whether explicitly or implicitly, on two kinds of mathematical infrastructure: a theory of how heterogeneous information can be aggregated into a single representative value, and a topological framework that guarantees the orderly operations defined over fuzzy sets. Chapters 10 and 11 provide this infrastructure directly.

Chapter 10, by Blanco-Mesa, constitutes the theoretical nucleus of the collection. The chapter proposes novel heterogeneous penalty-based aggregation functions, including the Penalty Weight Adequacy Coefficient (P-WAC), its double-penalty variant (dP-WAC), and the Bonferroni Penalty Adequacy Coefficient OWA (BPAC-OWA). These operators extend the Bonferroni mean and the Ordered Weighted Average family by incorporating the adequacy coefficient as a penalty mechanism, enabling decision makers to model disagreement, asymmetry, and heterogeneity in expert information within a single unified formula. The mathematical properties established in the chapter, including generalized and quasi-arithmetic extensions, represent more than a technical exercise, as they provide the formal grammar through which the applied chapters can capture the complexity of real decision environments.

 DOI: 10.1201/9781003724902-13

The operators proposed here can be understood as a new generation of aggregation tools specifically designed for scenarios where expert opinions diverge and where the structure of that divergence is informative.

Chapter 11, by Likhada, complements this aggregation infrastructure at a deeper mathematical level by introducing new categories of compactness in fuzzy topological spaces: fuzzy βSC-compact, fuzzy βC-compact, and fuzzy β-almost compact spaces. The chapter establishes their interrelations, provides counterexamples that delimit the boundaries of each concept, and positions these new categories within the existing hierarchy of fuzzy compact spaces. The aggregation operators of Chapter 10 and the fuzzy decision methods applied throughout the book operate within spaces defined by fuzzy topological structures. The compactness properties established in Chapter 11 guarantee that certain convergence and covering behaviors hold in these spaces, ensuring that the optimization procedures, limit operations, and continuity arguments upon which fuzzy decision methods rely are mathematically well founded. In this sense, Chapter 11 provides the foundational layer upon which the entire edifice of fuzzy decision-making rests, and its contribution to extending the theory of fuzzy compactness opens new avenues for the development of decision models in engineering, biomedical applications, and design science.

C.2 FROM THEORY TO COMPUTATIONAL ARCHITECTURE: NEUTROSOPHIC DECISION SYSTEMS

The transition from mathematical foundations to operational decision systems requires intermediate constructs that translate theoretical properties into implementable computational frameworks. Chapter 9, by Bardhan and colleagues, occupies this bridging position by integrating neutrosophic logic with blockchain technology.

Neutrosophic sets extend fuzzy and intuitionistic fuzzy frameworks by explicitly incorporating indeterminacy as an independent component alongside truth and falsity. This representational richness is particularly valuable in distributed decision environments where information is not merely imprecise but may be incomplete, contradictory, or actively contested. The Neutrosophic Blockchain Framework proposed in Chapter 9 leverages the decentralization, immutability, and transparency of blockchain to create a decision architecture in which neutrosophic evaluations are recorded, aggregated through a neutrosophic consensus mechanism, and executed through smart contracts. The three case studies presented in supply chain management, healthcare diagnosis, and financial risk assessment demonstrate that this architecture presents robustness and adaptability when information is incomplete or contradictory.

The placement of this contribution within the volume is strategic. Where Chapter 10 provides the aggregation operators and Chapter 11 provides the topological guarantees, Chapter 9 demonstrates how uncertainty-aware mathematical constructs can be embedded within a computational infrastructure that ensures auditability, traceability, and distributed trust. This progression from mathematical theory through computational architecture to applied deployment defines the methodological progression of the book.

C.3 MATHEMATICAL DECISION METHODS ACROSS DOMAINS: APPLIED DEMONSTRATIONS

Five chapters apply established and adapted MCDM frameworks to decision problems drawn from finance, organizational management, public policy, supply chain engineering, and tourism governance. Each instantiates a common analytical logic: identify a decision problem structured around multiple competing criteria, model the uncertainty inherent in expert judgment or observational data, apply an aggregation or outranking procedure appropriate to the structure of the problem, and extract a ranking or classification that supports actionable recommendations. Yet the specific choices each chapter makes regarding method selection, criterion definition, and domain adaptation reveal the flexibility and context sensitivity that distinguish wide-ranging applications of mathematical decision science from mechanical formula application.

C.3.1 Financial Markets and Portfolio Construction

Chapter 2, by Muñoz Palma and colleagues, demonstrates the power of the Multiple Criteria Hierarchical Process (MCHP) combined with ELECTRE III for the selection of Exchange-Traded Funds. By structuring criteria into a two-level hierarchy of market risk and financial indicators, the approach captures the interaction between risk-oriented and return-oriented sub-criteria in a way that flat multicriteria methods cannot. The analysis of 20 global ETF assets yields a portfolio-ready ranking responsive to investor risk profiles, illustrating how hierarchical criterion structures can reveal preference patterns that single-level evaluations obscure.

C.3.2 Organizational Governance and Technology Alignment

Chapter 3, by Parroquín-Amaya and colleagues, applies Fuzzy TOPSIS integrated with Principal Component Analysis to the prioritization of 68 critical success factors for IT Governance implementation. The hybrid approach is distinguished for its methodological sophistication: PCA identifies the latent dimensional structure of the governance problem, reducing complexity without discarding information, while Fuzzy TOPSIS incorporates the irreducible subjectivity of expert judgments into the ranking process. The resulting six-dimensional framework provides IT managers with a structured and prioritized tool that connects technological investment to strategic objectives, demonstrating that fuzzy multicriteria methods can serve as bridges between the technical specificity of IT systems and the strategic generality of organizational goals.

C.3.3 Public Policy and Social Entrepreneurship

Chapter 4, by Valadez-Solana and colleagues, applies the Ordered Weighted Average (OWA) operator to rank Mexican state governments by their capacity for social entrepreneurship. The sensitivity analysis conducted across five aggregation operators reveals that the ranking is robust for high-performing states, with Estado de

México consistently leading, while exhibiting meaningful variation in the middle and lower tiers. This finding is analytically significant because it demonstrates a property of aggregation operators that the theoretical literature has long emphasized but that applied studies rarely illustrate with such clarity: the choice of aggregation strategy is consequential, and different weighting vectors produce different orderings that carry different policy implications. The methodological transparency with which this sensitivity is reported is itself a contribution to evidence-based public policy and a model for how aggregation results should be communicated to non-technical decision makers.

C.3.4 Supply Chain Resilience and Behavioral Decision Science

Chapter 5, by Holguín Ávila, integrates the TODIM method with digital twin technology for supply chain performance assessment. TODIM is grounded in prospect theory and is therefore sensitive to risk perception and loss aversion, behavioral dimensions that classical optimization approaches assume away but that are pervasive in real procurement and logistics decisions. The chapter compares TODIM with alternative approaches such as AHP and PSI, articulating why a behaviorally realistic method is preferable for supply chain decisions characterized by asymmetric risk preferences. The conceptual integration with digital twin technology points toward a future in which multicriteria evaluation is embedded within continuously updated simulation environments, enabling proactive rather than reactive decision making.

C.3.5 Tourism Governance and Indicator Development

Chapter 8, by Herrera-Prado and colleagues, applies the Analytic Hierarchy Process to the development of the Assessment Model for ICT Indicators (AMII) for Smart Tourism Destinations. By integrating evidence from four established STD conceptual models and eliciting expert-weighted priorities across five strategic axes of Governance, Accessibility, Sustainability, Innovation, and Technology, the AMII provides a locally adaptable, certification-compatible evaluation tool. The chapter exemplifies a function of MCDM that extends beyond ranking: AHP serves here as a participatory instrument for operationalizing contested theoretical frameworks into measurable indicators, a function whose methodological value extends well beyond the tourism domain to any field in which abstract policy concepts must be translated into concrete evaluation criteria.

The complementarity among these five applied chapters reflects a wide-ranging pattern in the development of mathematical decision science. Governance capacity and operational resilience, as demonstrated by Chapters 3 and 5, respectively, are increasingly recognized as attached, and MCDM methods that incorporate behavioral realism are better suited to contemporary organizational environments than approaches that assume fully rational, fully informed decision makers. The sensitivity analyses of Chapter 4 and the hierarchical criterion structures of Chapters 2 and 8 illustrate that the selection and configuration of aggregation operators is itself a substantive decision, one that shapes outcomes and should be subjected to the same rigor as the primary analysis.

C.4 EVALUATION AT SCALE AND THE MAPPING OF RESEARCH FRONTIERS

Two chapters in the volume serve a distinctive function that is essential to the intellectual completeness of the collection. Rather than applying MCDM methods to solve a specific decision problem, Chapters 6 and 7 operate at the level of large-scale empirical evaluation and systematic knowledge mapping, respectively. Their contribution to the volume is best understood not as the deployment of fuzzy or multicriteria tools but as the identification of the domains and conditions under which such tools are most needed and least developed.

C.4.1 Large-Scale Quantitative Evaluation as a Foundation for Decision Modeling

Chapter 6, by López-Barraza and colleagues, presents a multi-dimensional evaluation of student satisfaction in Mexican upper secondary education institutions. With over 11,000 responses across 25 indicators and a margin of error below 1% at 99% confidence, the study demonstrates the feasibility and value of systematic, large-scale satisfaction measurement at an educational level where such instruments have been notably absent. The chapter reveals, with considerable statistical precision, the dimensions along which student experience varies and identifies targeted improvement priorities. From the perspective of the volume's overarching argument, the significance of this contribution is twofold. First, it provides the kind of rich, multidimensional empirical base upon which future multicriteria models can be constructed: the 25 indicators evaluated could serve as criteria within a Fuzzy TOPSIS, OWA, or ELECTRE framework, weighted by institutional priorities and aggregated according to the decision maker's risk posture. Second, it illustrates the limitations of purely statistical approaches that treat all dimensions as equally important or that reduce satisfaction to a single composite score. The contrast between what Chapter 6 achieves through quantitative evaluation and what the MCDM methods of other chapters could additionally reveal through weighted aggregation and sensitivity analysis defines a productive research frontier at the intersection of educational measurement and mathematical decision science.

C.4.2 Bibliometric Mapping as the Identification of Methodological Demand

Chapter 7, by García-Orozco and colleagues, provides a comprehensive bibliometric mapping of the scientific literature linking food insecurity with diet-related noncommunicable diseases including diabetes, obesity, and hypertension. The analysis covers 3,650 documents from 1,178 sources, revealing a sustained annual growth rate exceeding 20% and a field that has consolidated around a stable conceptual core of access pathways, metabolic risk, and health disparities. While bibliometric analysis is methodologically distinct from MCDM, its function within this volume is strategically significant. The chapter performs a systematic aggregation of heterogeneous evidence across thousands of documents, identifies the thematic dimensions driving the field's intellectual structure, and reveals unresolved research priorities with precision. More importantly, the chapter's own findings point directly toward the methodological toolkit that the rest of the volume develops. The reviewed literature is characterized by the predominance of observational and epidemiological approaches with limited integration of adaptive models, decision-support systems, and analytical tools capable of handling complexity and territorial heterogeneity. The chapter explicitly identifies the integration of intelligent systems, fuzzy logic, and decision-support methodologies as a promising avenue for advancing the field. In this sense, Chapter 7 functions as a compass within the volume: it maps a rapidly growing, policy-relevant research domain where the mathematical and fuzzy decision tools developed in Chapters 2 through 5 and 8 through 10 have not yet been systematically deployed, thereby identifying the demand for the very methodologies the book advances. The opportunity to apply multicriteria frameworks to the prioritization of food security interventions, the evaluation of competing nutrition policies, or the assessment of regional vulnerability to diet-related disease represents a natural and productive extension of the research programmed this volume inaugurates.

C.5 THE ARCHITECTURE OF THE VOLUME: CONVERGENT INSIGHTS

Reading the ten chapters as an integrated argument rather than as independent contributions reveals a set of convergent insights that constitute the volume's collective contribution to mathematical and fuzzy decision science.

C.5.1 The Epistemological Primacy of Uncertainty

Every chapter confronts data or judgments that are imprecise, incomplete, or contested. The theoretical chapters respond through formal extensions of fuzzy logic, neutrosophic reasoning, and aggregation theory. The applied chapters respond through the deliberate selection of methods designed to function reliably precisely when uncertainty is inherent. The evaluative chapters reveal fields in which uncertainty is universal but the tools to handle it rigorously are not yet in place. This convergence is not accidental; it reflects the volume's deeper argument that the aspiration to reduce all decision inputs to crisp, deterministic values is not only unrealistic but analytically distorting, and those methods which honor uncertainty as a constitutive feature of complex problems are both more intellectually honest, practically useful and humanist.

C.5.2 The Consequentiality of Aggregation Choices

A recurrent finding across the applied chapters is that the selection of aggregation strategy materially affects the resulting rankings and recommendations. The sensitivity analysis of Chapter 4, the hierarchical distillation of Chapter 2, and the prospect-theoretic weighting of Chapter 5 each demonstrate that different operators encode different assumptions about the relative importance of criteria, the treatment of outliers, and the attitude toward risk. The penalty-based operators of Chapter 10 formalize this insight by making disagreement and deviation explicit components of the aggregation process. Collectively, the chapters establish that aggregation is not a neutral computational step but a substantive decision that shapes outcomes and should be reported with full transparency.

C.5.3 The Progression From Tool to Methodology to Institutional Practice

Several chapters move beyond the application of a single tool to construct integrated methodological frameworks adapted to institutional contexts. Chapter 3 combines Fuzzy TOPSIS with PCA to produce a governance decision model. Chapter 5 integrates TODIM with digital twin technology. Chapter 8 builds an indicator system through AHP combined with expert panel consultation. Chapter 9 embeds Neutrosophic logic within blockchain architecture, proposing the movement toward "how to construct a decision-support system that integrates mathematical rigor with domain knowledge and organizational constraints."

C.5.4 The Identification of Methodological Frontiers

The volume does not merely demonstrate existing capabilities, but it systematically identifies where the field's tools have not yet reached. Chapter 7 maps a major public health research domain where fuzzy and multicriteria methods are yet to be addressed. Chapter 6 reveals an educational evaluation context ripe for multicriteria enrichment. Chapter 11 opens new theoretical territory in fuzzy compactness that invites application. Chapter 10 proposes operators whose empirical deployment is explicitly reserved for future work. This dual character of the volume, simultaneously demonstrating what mathematical decision science can do and identifying where it has yet to go, positions it as both a consolidation of current knowledge and a prospectus for future research.

C.6 CLOSING REMARKS

Decision making under uncertainty is not a niche problem confined to a single discipline. It is the condition under which governments allocate resources, hospitals triage patients, firms evaluate suppliers, financial institutions construct portfolios, and educational systems assess their own effectiveness. The methods and applications assembled in this volume represent a research tradition whose practical relevance is growing as the decisions that shape societies become more complex, more consequential, and more dependent on the integration of heterogeneous information from multiple sources.

Aggregation operators that model disagreement and heterogeneity, outranking procedures that respect the incomparability of complex alternatives, fuzzy topological structures that guarantee the mathematical soundness of operations in uncertain spaces, and Neutrosophic frameworks that honor indeterminacy rather than eliminating it through forced precision collectively constitute a rigorous, flexible, and humanly realistic response to that condition. The bibliometric and evaluative chapters of the volume demonstrate that the demand for these tools is not hypothetical, but rather documented, growing, and in some cases urgent.

The ten chapters are, individually, contributions to their respective fields. Collectively, they advance a larger argument: that the development of formally rigorous, contextually sensitive, human-centered and practically actionable decision methodologies is a shared intellectual project that requires simultaneous progress in mathematical theory, computational implementation, empirical application, and the systematic identification of the frontiers where current tools fall short. Whether the reader approaches this volume from mathematics, management science, public health, tourism studies, engineering, or finance, the invitation is the same: to engage with that project and to carry it forward with the conviction that aggregation operators, fuzzy topologies, and neutrosophic architectures find their deepest justification not in formal elegance alone but in their capacity to support the irreducibly human act of deciding well under uncertainty, for the benefit of the people, institutions, and communities that those decisions ultimately shape.

Conclusions

Index